A Rational Energy Policy

A Rational Energy Policy

ROBERT J. BYER

AUGUST 2011

Library of Congress Control Number: 2011914918

ISBN:	Hardcover	978-1-4653-5564-5
	Softcover	978-1-4653-5563-8
	Ebook	978-1-4653-5565-2

This book was printed in the United States of America.

To order additional copies of this book, contact:
Xlibris Corporation
1-888-795-4274
www.Xlibris.com
Orders@Xlibris.com
103789

CONTENTS

ACKNOWLEDGEMENTS

I OWE A debt of gratitude to the several people who have reviewed the drafts of the book and provided useful suggestions to improve it. Special thanks are owed to James Roney whose PhD in engineering and experience in the nuclear, wind and solar energy fields were most useful. He provided some of the content dealing with wind and solar electric power generation as well as a check on my work. He assures me that I said nothing foolish, misleading or inaccurate relative to the several technologies mentioned in the text.

Bob Byer

This book outlines an energy policy for the United States which offers:

Energy Independence:

Elimination of the $1,000,000,000 per day paid for foreign oil purchases.

Environmental Improvements:

Elimination of the emissions and greenhouse gases from most transportation and power generation activities.

Economic Development:

Expansion of oil and natural gas industries with significant growth in jobs, low cost power to support manufacturing industries, and increased government revenues.

PREFACE

I HAVE WRITTEN THIS book because the US does not have, nor has it had, an energy policy for too long a time. Every effort to create one has been stymied by special interest groups with strong biases and deep pockets. Here we are, almost forty years since the first OPEC boycott in 1973 and we are still diddling around the edges of our energy problems.

In lieu of a well thought-out policy with explicitly defined objectives we have ended up with a multiyear string of legislation and regulation focused on subsets of our energy problems. The legislation often ignores the bigger picture of having sufficient energy to meet our needs. We require low cost power for transportation fuels, and low cost electric power to sustain and grow our economy. True, the air we breathe and the water in our rivers and streams are cleaner than they were years ago; still our efforts lack focus and positioning into a cohesive set of actions leading to fulfillment of defined objectives.

Our objectives, very simply ought to be:

- Energy Independence
- Cleaner Air
- Economic Expansion
- Strengthen our Economy
- National Security

These objectives, further defined in chapter 1, will take quite a while to accomplish, but are attainable if we individually and collectively tell our government what we, the citizens, want. The government needs to remove the many impediments that it has created. It must also give priority and restructuring of government operations to facilitate and speed up government review, licensing and permitting of projects whenever relevant.

The US is capable of doing great things involving huge engineering challenges within short schedules. It has done so before. Cases in point: The Manhattan Project in early 1940s, the Strategic Submarine and ICBM development project in the late 1950s and the development of the capability to put men on the moon in the 1960s. All were successful. We can do it again. It is vital that we get at it soon.

Our economy has been floundering. Unemployment is much too high. Adoption of the Rational Energy Policy can give the economy a greatly needed boost. Read this book; then contact the White House, and your Senators and Representatives in Congress. Contact your Governor's office and elected state officials. They need to push the federal government and insist your state gets a share of relevant royalties from extracted oil and gas. Everyone should insist that the federal government needs to stop dithering and focus on adopting a Rational Energy Policy. Our future depends on it.

CHAPTER 1

A Rational Energy Policy

Introduction and Objectives

ENERGY INDEPENDENCE HAS been mentioned in the press and the Congress for many years. Congress has occasionally passed legislation which purports to be doing something in the name of energy independence. We have ended up with a Department of Energy, standards for automobile mileage, the Energy Star program and more efficient appliances around our homes. Our manufacturers have also designed and installed more efficient manufacturing processes and so we are much better off—or are we? Even today in 2011, Congress is still spawning legislation in the name of energy independence, but they are still only tinkering on the edges. Nothing they do confronts the problem head-on in any significant way. Nothing they seem willing to do will dramatically reduce our oil imports, especially in the face of continuing population growth and our requirements for more cars and trucks and for more electric power. In this book we'll show you that we have plenty of oil and the other resources. We need, however, to become energy independent in a relatively few years and reap the many economic benefits that are equally as important as energy independence. Achieving this will require an immense attitudinal change, as well as changes in the way the government manages our energy resources. It will also require policy decisions by the President and the Congress. They must recognize we have the resources at hand to greatly improve our lot by becoming proactive in making the changes in public policy and law needed to facilitate meeting the objectives enumerated.

The objectives of our Rational Energy Policy are:

1. Energy independence: Minimization and ultimate elimination of the over $1 billion per day we spend for foreign oil.
2. Cleaner environment: Minimization and ultimate elimination of manmade noxious and unnecessary greenhouse gases from transportation and power generation activities.
3. Economic expansion: Significant creation of permanent jobs.
4. Strengthen our economy: Growing government revenues enabling officials to pay down the national debt and reduce the trade deficit.
5. National Security: Increase our national security by minimizing dependency on foreign oil and possibly minimizing the size of our standing military forces.

We can accomplish all of these with the policies advocated.

This book provides data to illustrate how history has brought us to the point where we need an energy policy to guide us to fulfillment of these objectives.

The Growth of Energy Dependence

While the Congress does well from time-to-time, they also issued drilling moratoria on Federal lands and most places in our coastal waters which prevent our use of available energy resources. Clearly the Government is on the one hand calling for energy independence, while on the other hand protecting the environment to a degree which has brought us farther from energy independence than we were in 1973. It sounds like we are conflicted and unable to arrive at a balance that will provide both energy independence and the protection of our environment that we all want.

So how far have we come in achieving "energy independence" in the 35 years since the OPEC oil embargo of 1973. The US Department of Energy, Energy Information Administration (EIA) provides us with the 2008 summary data shown in Table 1-1 below: The figures speak for themselves; declines in domestic production, and big increases in oil imports

Table 1-1 Petroleum Overview 1975-2008			
	1975	2008	Change
Domestic Production of Oil			
Barrels	3,652,737,000	2,464,620,000	
Barrels per Day	10,007,498	6,752,382	(32.5%)
Net Imports			
Barrels	2,133,907,000	4,040,951,000	
Barrels per Day	5,846,320	11,071,098	89.4%
Petroleum Products Supplied[1]			
Barrels	5,957,515,000	7,107,242,000	
Barrels per Day	16,321,958	19,471,895	19.3%
Percentage of Oil Usage that is Imported	**36**	**57**	**58%**

Next, we provide a quick look at the sources of our energy in 1975 and 2008[2]. Table 1-2 below provides data categorized as fossil fuels (coal, natural gas and oil), nuclear power, and renewable energy (wind, solar, and biomass). The energy data are provided in British thermal units (BTU) to facilitate comparison.

Table 1-2 Energy Sources and Consumption 1975-2008		
Sources	Quadrillion BTUs	
	1975	2008
Production		
Fossil Fuels	54.733	57.940
Nuclear Power	1.900	8.455
Renewable Energy	4.723	7.316
Total Domestic Energy Production	61.357	73.711
Net Imports (petroleum and natural gas)	11.703	25.775
Total Production and Imports	73.060	99.486
Consumption		
Fossil Fuels	65.335	83.436
Nuclear Power	1.900	8.455
Renewable Energy	4.723	7.316
Total Energy Consumption	71.999	99.304

1 The petroleum products supplied data in Table 1-1reflects some adjustment for changes in the Strategic Petroleum Reserve and other situations.

2 1975 is approximately the year we started talking about energy independence.

Table 1-3 contains data on the cost of imported oil over the last few years. The data is all from the DOE's Energy Information Agency

Table 1-3 Oil Import Data				
Year	Barrels Imported Million	Average $ per Barrel	Total Cost Per Year $ Billion	Average Cost Per Day $ Millions
2007	4,916	$66.36	$326	894
2008	4,727	$90.32	$427	1,170
2009	4,267	$57.78	$247	675
2010	4,290	$74.20	$318	872

These three tables tell us the following:

- First the imbalance between energy independence and protection of the environment now requires us to import approximately 11,000,000 barrels of oil every day at a cost of approximately $900 million to $1 billion dollars per day depending on the fluctuating price of oil.[3]
- Second, these numbers have been growing since the early 1970s when the first calls for energy independence were sounded. Both nuclear energy and use of renewable energy have increased but not nearly enough to keep up with the growth in total consumption. Unfortunately the consumption of fossil fuels, the sources that give us the most environmental problems, has been growing the fastest in absolute terms.

The cost of petroleum imports is unsustainable. The time to do something about this is now.

We are in the worst recession since the great depression of the 1930s. The government's official unemployment rate is over 9% and when people no longer looking for work are counted, the unemployed number is 18 % or more. It doesn't need to be this way. A national program directed to energy independence and decreasing use of fossil fuels would be beneficial. It would put a large number of people back to work, increase our gross

---

3 The average daily cost of Imports peaked in 2008 and declined in 2009 due to the recession. The average daily cost resumed the upward trend in 2010 and will be back to a billion dollars or more in 2011.

domestic product and eventually provide the inexpensive power we need to facilitate growth in manufacturing. Most importantly, we also believe the strategy we define will ultimately lead us to a cleaner and healthier environment.

The growth in GDP would increase tax revenues at all levels of government and place us in a better position to handle our national debt. It would also restore a more promising future for our young people. It is difficult to predict the extent of economic growth that would accrue from not spending the millions every day now spent to buy foreign oil, and instead spend the funds to develop our own energy resources. But the direct and indirect employment and other economic impacts would be substantial. Some estimates of economic impacts are included later in this book.

Think about what the cost of buying foreign oil really means to us. If we produced all of that oil in the United States how many people would the oil industry need to do so? In addition, all of the new hires would spend their wages buying goods and services locally resulting in the indirect employment of many more people. When these new direct and indirect employees pay taxes our government's (local, state and national) would be far better off than they are today. Maybe even our exploding debt situation could become manageable. You might say we don't have the oil which is why we buy so much. We will show you in this book that we have far more oil than you might have thought and that energy independence is not only desirable but imperative. We can become energy independent, but some changes may be required, and no great sacrifices are needed. In the end, our objective is to develop our own energy resources with our own people so as to improve our employment and economic condition. For purposes of our analysis, we look at energy as that which is required for transportation purposes and electric power generation.

In this book we will summarize the energy situation today, and inform you of some of the relevant things going on in the US that influences our strategy for proceeding. In addition, we offer a suggestion for the States to pursue to help them achieve benefits from activities which will bring us energy independence. You should note the failure to attain energy independence over the last 45 years can be laid at the feet of our political leadership; specifically their lack of leadership. We decided to write this book because our politicians need to hear from us that we want a comprehensive program for energy independence ASAP, with regular progress reports to the nation. Our readers will be informed so they can tell the people, who work for us in Washington, that its time they acted to bring us real energy independence. A later chapter of this book outlines what they have to accomplish to make it happen. Read it and then contact your representatives and senators and by all means, the White House. Tell them: No procrastination permitted. Do it now!

We believe that a comprehensive program could be designed to get us to real energy independence in a few years. In doing so we will lead our country out of the economic doldrums we are now experiencing. We also believe that we can get Congress to remove the impediments to energy independence if they hear from many of their constituents demanding the actions required to achieve this goal. We list the impediments and recommended actions in Chapters 8 and 9.

Elimination of Emissions is an Important Objective

Our government chose moratoria on drilling on US territory as a way to preserve the environment by avoiding accidental spills and whatever damage to the landscape or the ocean that might accrue from drilling. Perhaps damage accrued often enough to justify the decisions to issue moratoria and pay the world price for oil. All of this started when oil was only a few dollars per barrel. The passage of time, however, has seen much improved awareness of the environment and the need to protect it. This has manifested itself in workplace rules and regulations involving safety and environmental protection. Employees generally appreciate the need to follow the rules and improved compliance has resulted.[4] Over the years since the first oil drilling moratorium was imposed, several important factors have changed the complexion of what we were doing. We declared more moratoria, our need for oil increased dramatically, as did the price of oil. What once had little economic impact has become very expensive in terms of the daily cost of oil purchases. We can now anticipate that drilling

---

4 Accidents will occasionally happen, however rarely. Note there were literally thousands of wells drilled in the Gulf of Mexico without environmental problems before we had the accident in the summer of 2010. It was an expensive accident but it doesn't warrant any need to curtail exploitation of the energy resources in the Gulf.

in the US is safe and generally environmentally clean. Accidents may happen from time to time but these can turn-out to be learning experiences that may prevent future mishaps. Conclusion: Environmental protection remains in the national interest but the benefits of drilling in our country can outweigh the occasional oil spill. We can now safely avoid the cost of paying for foreign oil. We'll address other benefits later in this book.

A rational energy policy has to deal with air polluted by a hundreds of utilities and about 220,000,000 cars and trucks. The pollution from autos and trucks can ultimately be dealt with by focusing research and development processes on electric propulsion systems although we have a long way to go as we discuss in Chapter 2. The pollution from power plants is a huge problem without easy solutions. The government is currently trying to define new regulations for elimination of emissions including greenhouse gases. Unfortunately the equipment and processes the utilities will need to comply with these regulations is expensive and adds to the cost of power that will be passed along to customers. The result will make residential customers unhappy but more importantly manufacturers who need low cost power to be competitive in their industries may find the increased costs unacceptable. They might decide to go elsewhere to find cheaper power. Consequently we need to look to other alternatives for power that are cleaner and less expensive.

Technology Development Impedes Attainment of Our Ultimate Objectives

We will show that we will have to drill for oil and natural gas for several more years until an engineering break-through is achieved primarily related to battery improvements for cars. In developing the strategy to achieve energy independence we were guided by several major requirements. We must end up being as close to totally independent as possible allowing only imports of selected raw materials.

We must minimize environmental impacts and ultimately, over a period of time, arrive at a point where all cars and many trucks are powered by electricity.

Electric power plants will be mainly nuclear powered that do not emit pollutants into the environment. In addition we will have solar and wind power in locations where they can be justified. Since replacing dirty power plants with environmentally clean ones takes a long time, we must regard natural gas for the electric utilities as an interim fuel. Natural gas is much cleaner than coal. Conversion costs to natural gas can be rationalized on an economic basis because the process of converting to nuclear plants may take years.

Along the way to an all electric economy we will have to accept a gradual evolution where we will see improvements in batteries enabling better hybrid vehicles, conversion of some vehicles to natural gas, improved gasoline engines and fuel cell powered cars. These will take years to accomplish so in the meantime our needs for oil based fuels will continue. But hopefully in diminishing amounts.

In the end, we want no imports from countries whose friendship is doubtful. We want to move to an economy based on domestic non-polluting energy sources and low cost electricity to support economic development, high employment and growth.

If you have read this far and call yourself an environmentalist, you may be asking why should we accept more drilling when it costs us a billion dollars a day—small price for clean water, clean coasts, and clean air? Maybe you should consider that there might be additional dimensions to the price. For example, economic development **not** happening, jobs and income **not** created, taxes **not** paid at all levels of government; all from growth **not** happening because we are **not** expanding our drilling industry when we send our money to foreign oil suppliers.

The Organization of This Book

The chapters of this book are organized to present our story as follows:

Chapter 2 looks at transportation uses of oil and its derivatives, namely gasoline and diesel fuel. This requires about 14 million barrels per day. We look at the efforts to develop alternative fuels, hybrid cars and eventually electric cars. We'll show why this is unlikely to reduce our oil requirements any time soon concluding with an understanding that if we want to reduce foreign imports we are driven to increase domestic oil production.

Chapters 3 and 4 focus on electric power generation identifying the various fuels in current use, the problems and benefits associated with each. An interesting fact is that oil is a minor player in this arena. Our focus is on coal, natural gas and the three that we need to rely upon, nuclear, wind and solar. Hydropower has largely been developed with little remaining for expansion. Similarly geothermal resources are limited in quantity. These two resources are therefore not further addressed in this book.

Chapter 5 provides useful information about the electrical distribution grid and address requirements to modernize it to increases its' reliability and efficiency.

Chapter 6 addresses the availability of domestic oil resources. If you examine the government Energy Information Agency's resource availability data you might quickly conclude we buy foreign oil because we lack domestic resources. Such is not the case. The government only shows the information for legally available oil. The energy resources where access is denied because of government imposed moratoria are hidden. It requires considerable effort to develop more inclusive estimates of our energy resources. We have searched for and found data you will find interesting and provocative.

Chapter 7 Obtaining more oil from domestic source and modernizing our electricity generation and distribution infrastructure is essential to meet the demands of our ever expanding population and provide the basis for a robust economy. Expanded drilling in areas now closed by moratoria, converting old coal fueled power plants to natural gas, or building new nuclear power plants, may have a profound impact on our economy. Doing it right by focusing on clean low cost power will support manufacturing industries and provide jobs for years to come. We look at several studies that revealed employment and other economic impacts. While some of these have specific geographic focus, they suggest a nationwide view could have immense economic impact.

Chapter 8 presents our recommendations which collectively comprise the Rational Energy Policy. It's includes components for oil drilling and electrical power generation, the grid for electrical power distribution, and basic research where the government has been and should continue to make a contribution. A recommendation is provided indicating how the states could make a contribution to relieving our demand for oil. In the process they can facilitate the conversion of commuter's cars to natural gas powered vehicles.

Chapter 9. The government, for whatever reason, has created impediments that have spawned our energy problems ranging from restricting vast areas of our country from exploiting our energy resources to creating bureaucracies that tightly control drilling for oil or permitting electrical power plant development. We expose these impediments and suggest that sympathetic readers contact relevant politicians in Washington to seek acceptance and implementation of the Rational Energy Policy.

CHAPTER 2

Fuel for Transportation Purposes

Introduction

CHAPTER 1 PROVIDED data on our energy imports. This chapter provides information to help you understand how much of our use of energy is for transportation purposes and why turning off the oil import spigot will not happen quickly. The slow change in the inventory of cars and trucks limits the rate at which new regulation or fuel economizing technology improvements can be introduced. Finally, government imposed impediments to drilling and inertia standing in the way will delay the achievement of significant results. The longer we wait, however to initiate removal of impediments so we can increase drilling in the US, the longer it will take to achieve significant results and the longer it will take to ramp up our economy.

We'll start by looking at what we do with each barrel of oil, all 19,500,000 of them every day.

The DOE's Energy Information Agency (EIA) provides us with the data in Table 2-1 below:

Table 2-1 Products Made From a Barrel of Oil		
Product	**Gallons of Product**	**Percent**
Gasoline	19.15	
Diesel Fuel	9.21	
Jet Fuel	3.82	
Total Transportation Related Products	**32.18**	**72**
Other Distillates (heating oil)	1.75	
Heavy Fuel Oil	1.76	
Liquefied Petroleum Gases	1.72	
Other Products	7.27	
Total Non Transportation Related Products	**12.50**	**28**
Total	**44.68**[5]	**100**

We clearly see that roughly 72 percent of our oil usage is required for transportation purposes-cars, trucks, airplanes and railroads. Seventy two percent of the roughly 19.5 million barrels of oil per day is 14 million barrels per day for transportation purposes. If we could significantly reduce this we may be able to entirely eliminate oil imports. This then is the focus of our attention in this chapter. [We'll ignore the remaining 28 % of our usage as some of this is heating oil and feedstock for various industries, plastics and the like.]

While a large amount of the fuel used for transportation involves the railroads and aircraft, the biggest portion is for cars and trucks. The auto and truck registration data helps us to appreciate how big a problem this is. Table 2-2 consists of US Department of Transportation vehicle registrations data.

[5] A 42-U.S. gallon barrel of crude oil yields between 44 and 45 gallons of petroleum products. These totals are greater than 42 gallons due to processing gain.

Table 2-2 Vehicle Registrations 1975 and 2000					
	1975	2000	Change	Approximate Vehicle Annual Replacements	Time to Replace the Entire Fleet with More Efficient Vehicles
Automobiles	106,706,000	133,121,000	25%	10,000,000	13 + years
Trucks and Busses	26,243,000	87,854,000	235%	6,000,000	14 + years
Total	**132,949,000**	**221,475,000**	**67%**		

This data clearly indicates that actions taken to increase vehicle fuel economy will require possibly 13 or more years just to convert the inventory to new and more fuel efficient engines. The 13 plus years cannot start until updated technology or new fuel is available for commercialization. For example, recent increases in automobile CAFÉ standards will require two to four years for the automobile industry to design compliant vehicles and followed by another thirteen or more years to get all of the older, less efficient cars off the road.

New Technologies are Evolving to Power Cars and Trucks and Relieve Dependence on Oil

At this time we don't know how our cars and trucks will be powered in the future. There are certainly a lot of candidates including biomass, hybrid electric, all electric, natural gas and others. There are problems with each of these possibilities that we address below. Each of them, however, as they are readied for commercialization will take years to establish significant positions in the inventory of registered vehicles. Consequently one can only conclude that we will be dependent on gasoline for many years to come. The big question then is, will we buy the oil in the world markets or will we drill for it at home and employ our people in the process?

Biomass

Biomass could be a significant contributor to the solution of fuels problem. The primary candidates are corned based ethanol and cellulosic ethanol. Corn based ethanol has been around for years and some diesel fuel is made from waste cooking oils. Neither of these however, has yet achieved significance in solving our energy problems. This is evident in the data shown in Table 2-3. The table provides EIA's Bio Fuel Production data. It is hard to imagine that biomass could ever make a significant dent in our appetite for oil. For example, data in the table show fuel ethanol production at only 605,600 barrels per day and biodiesel at only 50,000 barrels per day both in their 5th year of production. These are small compared to the 14,000,000 barrels per day used for transportation purposes.

Cellulosic ethanol might someday change this. There is also considerable research and development underway to create fuel from algae using biogenetic engineering technology. Breakthroughs in technology are required, but any predictions of when these might be available and how effective they might be are highly speculative.

Table 2-3 Bio Fuel Production Data in Thousands of Barrels per Day					
Product	**2004**	**2005**	**2006**	**2007**	**2008**
Fuel Ethanol	221.5	254.7	318.6	425.4	605.6
Bio Diesel	1.8	5.9	16.3	32.0	50.5

It also has to be recognized that ethanol has less energy content than the gasoline it replaces when we fuel our vehicles. This data is shown in Table 2-4. It takes energy content to move our vehicles so less energy in a blend of gasoline and ethanol means we will obtain fewer miles per gallon. It also means we will use more fuel containing ethanol to drive equivalent miles.

Table 2-4 Energy Content of Fuels	
Product	**Energy Content BTUs/Gal**
Gasoline	**115,400**
Diesel Fuel	**128,700**
Ethanol	**76,000**
Compressed Natural Gas	**See note below**

Ethanol derived from corn is no different than ethanol derived from cellulosic materials. So we should strive to obtain the ethanol from the source that does this most efficiently. Both require energy to produce. It appears that cellulosic ethanol uses much less energy in its manufacture than does corn based ethanol and consequently it would be preferred. It is uncertain however when it will become an economically viable choice.

Note Table 2-4 does not provide data on natural gas which is also used as a fuel for trucks and cars. The omission is made because the energy content of a unit of the fuel depends on several variables including whether the gas is in a compressed form or liquid, and if compressed, at what pressure and the temperature. Honda Motors offers their Civic model in both CNG and gasoline powered. They claim the same EPA fuel economy for both models. This indicates that they have a CNG tank whose size is specifically calculated to provide equivalent performance.

Corn-Based Ethanol

Corn farmers have been raising corn to be fermented into ethanol for years. The ethanol has been blended with gasoline to help reduce oil consumption. The entire idea of diverting a food crop to supplement our energy supply has been criticized from the start as a bad idea. The case against it includes these factors: the amount of energy needed to raise the corn, transport it to the fermentation facility, and convert the corn to ethanol is, at best, marginally less than the energy in the ethanol created. Some studies have been done by people that conclude even more energy is required than is available in the finished product. On top of that the entire process is supported by government subsidies which add greatly to the cost. In recent times, as corn production has increased for ethanol production, US and world prices of corn have soared which has spread to other food products such as corn fed beef and so on. All in all, it appears raising corn to supplement our energy supply is a poor idea. And subsidizing it is a worse idea. When cellulosic ethanol is available the government should terminate all corn ethanol subsidies and save the taxpayers the money.

Cellulosic Based Ethanol

The manufacture of cellulosic ethanol is a totally different story. It is a story which hasn't yet reached a satisfactory conclusion. The difference between corn based ethanol and cellulosic based ethanol is that the former is developed from sugar and the latter from cellulose. Cellulose is widely available as plant and forest waste. While corn based ethanol requires the milling of the corn, addition of water, heat and enzymes to induce fermentation, the cellulosic process differs in a few ways; the heat used to make corn based ethanol is usually provided by natural gas. With cellulosic based ethanol two enzymes are required, one to breakdown the cell structure of the feedstock (waste) and the second to support the fermentation process. In breaking down the cell structure, some portion of the waste is burned to provide heat to support the fermentation to create the ethanol. This makes the cellulosic process far more energy efficient. Although carbon dioxide is created in burning waste, the amount created is balanced by the carbon dioxide which the plants absorb in the growth process. In the corn based ethanol manufacture there is a net increase in carbon dioxide from burning natural gas.

Ethanol, however, has a property that will forever inhibit its large scale usage and that is it is hydroscopic. It absorbs water. Pipelines which often bring gasoline to local markets don't allow the ethanol to be blended with gasoline before entering the pipelines. The water would facilitate corrosion in the pipe. Instead the ethanol is transported by tanker trucks and blended locally before arriving at service stations. The water does not appear to damage automobiles since it turns to steam in the car's engine.

Bio Diesel

Table 2-3 shows biodiesel production over a five year period. Note while production increases from year to year by 2008 it only reaches 50,500 barrels per day. This doesn't seem likely to ever make a significant contribution, since it is likely constrained by the supply of available waste vegetable oil and animal fat. All of the biodiesel is used by trucks so it will not help reduce automobile demands for fuel.

Hybrid Electric and Gasoline Cars

More and more auto manufacturers are making hybrid cars. Some use nickel-metal hydride and some use lithium ion batteries. All of the hybrids use a gasoline powered engine in conjunction with the electric motor. The battery is recharged whenever the gasoline engine is running and possibly by regenerative braking as well. The best of these hybrids is probably the Toyota Prius which was designed from the ground-up for high fuel economy (approximately 50 miles per gallon of gas). GM may be about to make a significant contribution to hybrid product offerings. It is expected to soon make its' Volt hybrid which is purported to have a small gasoline engine whose primary function is to charge a large battery. The battery will provide the primary motive force. This is unlike the Prius in which power to move the car is shared between the gasoline engine and the battery driven electric engine. Many hybrids are redesigns of standard models and the results are not usually very impressive. A common problem is that all of these hybrids are significantly more expensive than equivalent cars with only gasoline engines so unless you drive a whole lot, the fuel savings will require many years to justify the added cost of the hybrid. We don't yet know how consumers are going to react to these cars.

All Electric Powered Cars

Another problem with automobiles must be recognized. Hybrid automobiles, particularly with the anticipated "plug-in" hybrids which we are told may be recharged in our garages or publically accessible recharging stations are not a panacea. A number of problems may restrict the rate at which these vehicles will see the highway in significant numbers. First, we don't at this writing, have a working plug-in hybrid in cars nor do we have electric meters in our homes so we can recharge at night when power companies can sell power at lower cost because of reduced demand. The second problem is the cost of the batteries which is a large portion of the added cost of hybrid cars. Customers may decline to buy these cars unless there is a clear economic advantage for them to do so. A high priced battery and the need to replace it once in a while may eclipse the anticipated fuel savings. Thus anticipated acceptance of plug in hybrid cars may not be imminent until battery design and production technology evolve to improve the cost-benefit situation. [The incremental cost of a hybrid over a conventionally powered car must be less than the projected fuel savings over say the four or five years that a typical owner would keep a car.]

It should be noted however, batteries cannot be charged quickly. Thus recharging at a station functionally equivalent to a gas station may not be attractive. People will most likely want a quick refueling car which means the hybrid car may be the vehicle that will prevail 15 or so years into the future. The battery will be recharged by the gasoline portion of the hybrid propulsion system. An abundance of all electric vehicles may be in the more distant future.

One item is clear about electric cars. Among all the possibilities of internal combustion engines fueled by whatever evolves over the next few years, only the electric car has no emissions to pollute the environment. The sooner we reach the economically viable all electric car, the better off we will be. If we can concurrently convert to all nuclear power generated electricity we will also have reached a situation which is best for the environment and offers electricity at the lowest possible cost. Nuclear generated power is also the most stable; it's not impacted by the fluctuating world prices for oil. At that point we will have reached energy independence, much cleaner air and prices which are good for the economy.

Natural Gas as a Transportation Fuel

Natural gas has been sold as fuel for trucks and buses for years primarily in metropolitan areas to fleet operators such as power and bus companies. These vehicles are heavily used each day and return to a home base at the end of their workday. The site where these vehicles are stationed includes a refueling facility. Generally these fueling stations are not available to the public. They are owned by whoever owns the fleet. These owners might also use ordinary automobiles for some of their personnel. Therefore some auto manufacturers make natural gas powered versions of personnel cars. The public however, don't ordinarily buy these cars because they have no way to fuel them.

We show in Chapter Six that the United States has an abundance of natural gas. It's also a relatively clean fuel so it would seem to be a natural to replace gasoline as a vehicle fuel, especially since the technology is well developed. This may not be the case, however. There are several impediments to natural gas becoming more widely used as a transportation fuel. Different constituencies have different problems standing in the way of using natural gas as replacement for gasoline or diesel engines.

Long distance truckers: The American Trucking Association (ATA)[6] indicates that natural gas powered long-haul trucks sell at a significant premium to diesel powered trucks. In addition the fuel economy of natural gas powered trucks is inferior to the diesel equivalent. Natural gas can be sold in two different forms; compressed natural gas (CNG) and liquid natural gas (LNG). The truckers don't find either form acceptable. CNG has a lower energy density and consequently doesn't provide fuel sufficient

---

6 American Trucking Association *Is Natural Gas a Viable Alternative to Diesel for the Trucking Industry?* ATA web site June 2010

for heavy long distance trucks. The tanks to contain LNG would be too big and too heavy. LNG is prepared by lowering its' temperature to minus 260 degree F and might be acceptable but the lack of a refueling infrastructure is a big problem. Government subsidies to create an infrastructure might be a way around this impediment. There are, however, still other impediments. For example, according to the ATA, the prices of natural gas powered large trucks sell at a large premium to the equivalent diesel powered truck and have poorer fuel economy. The fuel tanks for LNG are double walled built like a thermos bottle to help keep the fuel cold and are consequently very heavy. The tanks can reduce the trucker's payload. To me this sounds like a tough sell. The tanks for CNG are lighter so this factor might be a smaller problem.

Local truckers: On the other hand, for heavy trucks like cement and municipal waste hauling trucks working in a local area, the CNG powered trucks may be OK and preferred in areas where air quality is a problem. CNG is a relatively clean fuel. Without assistance from the government to help build a CNG fueling infrastructure, this opportunity may fall by the wayside. If the companies building pick-up trucks and similar vehicles used by small businesses and craftsmen could provide CNG vehicles in an area with the fueling infrastructure, a significant market might be created to keep prices down. At that point people might find CNG powered trucks attractive and we could put a dent or two in the 87,000,000 registered trucks in the USA.

The driving public: As mentioned above there are no natural gas fueling stations either in cities and towns or along the intercity highways where they could refuel. Suppose however, the government helped create the fueling infrastructure in metropolitan areas mentioned above, the same infrastructure could service customers willing to buy CNG powered cars for commuting and similar use. It might even induce the car companies to create a car designed just for commuting and local use. Could this be a new market for the companies to exploit?

Fuel Cells for Vehicle Power

Fuel cells were invented years ago. While not new, their application to automotive uses requires further development which is on-going by several manufactures. Their attractiveness is due to several features. For example, they are fueled by hydrogen which combines with the oxygen in air to form water which is the only emission. Electricity is created by the fuel cell which drives the car's wheels. Another desirable feature of the cell is its simplicity. An organization called Automotive Fuel Cell Cooperation describes their cell this way: The fuel cell includes only three major parts, an anode, a cathode and a polymer electrolyte membrane. The hydrogen fuel is channeled through the anode while air is channeled through the cathode. The anode and cathode are separated by the electrolyte. A platinum catalyst in the anode splits the hydrogen into positive hydrogen ions and negatively charged electrons. The hydrogen ions pass through the electrolyte to the cathode while the negatively charged electrons are channeled onto a cable forming an electric current. The current powers the motors in a circuit ultimately leading back to the cathode. The electrons and positively charged hydrogen ions combine with the air to form water that is drained off. The simplicity of the device is beautiful but now the problems begin that have kept this invention off the roads.

The hydrogen fuel is not readily available like gasoline. The companies working to develop fuel cells for vehicle are reforming fossil fuels like gasoline or natural gas. Another device is put on the vehicle to handle the reforming. Reforming gasoline involves stripping the hydrogen out of the fossil fuel and releasing the carbon in the gasoline as carbon dioxide to the atmosphere. In doing so we are back to the objection held by believers that people are causing global warming with carbon dioxide emissions

There is an alternative to reforming fossil fuels and that involves using electrolysis to turn water into hydrogen and oxygen, No pipelines required. The electrolysis could be made wherever the hydrogen is needed. Were this to become widespread, massive amounts of electricity would be required pushing the electric utilities into major expansion programs.

Summary of Alternative Fuels for Cars and Trucks

The majority of our use of oil is for transportation purposes with most of that going for gasoline and diesel fuel. The bulk of the diesel fuel goes for trucks and railroad operations. Some is devoted to aircraft fuels but nothing needs to be done about this since the airlines can be expected to conserve fuel as much as possible for their own survival reasons.

The problem of reducing oil consumption with cars and trucks is the slow rate in which old vehicles are retired and replaced with vehicles with more efficient fuel economizing technologies.

A summary table of the preceding material relative to the alternative technologies propelling cars and trucks is provided in Appendix A

Potential Diesel Fuel Savings by the Railroads

Today, as we indicated, the US uses roughly 19,500,000 barrels of oil per day. Approximately 11,000,000 barrels of this is imported. In addition, the Association of American Railroads provides us with the following information on the use of diesel fuel by Class 1 railroads[7]. In 2007, 2008 and 2009, the Class 1 railroads moved, respectively, 1.771, 1.777 and 1.532 trillion ton-miles of freight. Table 2-5[8] shows railroad diesel fuel consumption in the same years.

Table 2-5 Diesel Fuel Consumption of US Railroads			
Millions of Gallons			
2006	2007	2008	3 Year Average
4,192	4,062	3,886	4,048

The average of 4,048 million gallons of diesel fuel is equivalent to 96 million barrels of diesel per year or 263,000 barrels per day. Refineries general produce about 10 gallons of diesel from a 42 gallon barrel of oil so about a million barrels of oil per day may be required to produce this much diesel fuel. That's a lot of oil. Railroads are efficient movers of freight. Their most recent calculation of fuel usage is that a train can haul one ton of freight 457 miles on a single gallon of diesel fuel. The alternative fuel for the railroads should be electricity, as it is in Europe. Electrification of the railroads would not only allow us to reduce oil imports but it would also

7 Class 1 railroads include the seven largest US railroads.

8 1960-2008: Association of American Railroads, *Railroad Facts* (Washington, DC: Annual Issues).

eliminate emissions altogether except those from the power plants that supply the electricity. Our Rational Energy Policy advocates the ultimate conversion of most power plants to nuclear fuels. Nuclear plants are safe and do not emit any green house gases.

Electrification of railroads should be studied to determine the costs and benefits and to develop an implementation plan. If the plan were to proceed it should include electrification as well as improvements to the right-of-way that would permit high speed operations which in certain corridors would draw passengers off the air shuttles with further energy savings.

Aviation Fuel

There is little that can be done to minimize the use of aviation fuel that isn't already being done by the aircraft manufacturers, airlines and the FAA. These are highlighted below:

1. A big piece of the cost of running an airline is fuel costs. The airline industry is chronically in economic trouble therefore you can bet they are doing their best to minimize fuel costs.
2. In addition the air traffic control system is evolving into one of using technology to give pilots more control to fly routes that will minimize fuel usage, and finally,
3. The jet engine manufacturers have been working with airplane manufacturers, commercial and military, to make their products more fuel efficient. Coupled with the newer and lighter aircraft being built, particularly by Boeing, commercial aviation is on its way to being significantly more fuel efficient. So too is military aviation.

Replacement fuel for aircraft has high safety issues to hurdle.

Conclusion Relative to Reducing Oil Usage for Transportation Applications

Given the forgoing, we cannot expect the oil import situation to be any better in a few years unless we take direct action now. Our imports will increase and the cost will increase. Our only choices are to continue to plow ahead on the uncertain technology paths we are pursuing or take the certain path of increasing use of domestic resources. Only then, but probably not for a few years because of start-up lead times, will domestic oil production allow us to reduce imports. If we choose to exploit domestic energy resources, we increase employment, and improve the economy.

CHAPTER 3

Creating Electrical Power

THIS CHAPTER LOOKS at our options for generating electrical power. It addresses coal, natural gas and to a limited extent nuclear powered generating stations, and solar and wind generated electricity. These latter technologies are dealt with in greater depth in Chapter 4. We ignore hydroelectric power and some of the minor contributors that are unlikely to grow significantly. Hydropower is not minor but it is unlikely to grow since we don't have places to build more high dams. We are already fully exploiting the major sources of hydropower so there aren't significant opportunities to change the situation. In a later chapter, we will also look at the grid which transmits electrical power from the points of generation to the places where it is consumed.

Transportation activities, as we already discussed, are fueled mainly by oil thereby creating our oil import problem with the attendant high cost. Electrical power generation presents a different set of problems including negative environmental impacts, cost problems that impact consumers and the inefficient and sometimes unreliable electrical distribution system, the "grid". The emissions have to be eliminated because they cause health problems. The cost that manufacturers pay for power must be low to sustain manufacturing industries and support economic growth. Part of the cost users pay for power reflects energy wasted in the distribution system. Grid failures can be aggravating for everybody and they can be very expensive for businesses. Consequently we need to find a balance between available sources of energy that meets our objectives of no damaging emissions, and low cost and reliable power to support a robust and growing economy. We'll try to do that as we proceed.

Overview of Major Electricity Generation Sources

The really important factors when looking at policy alternatives for the fuels we use to generate power are, how much power we create with each fuel, the related emissions and resulting cost. Table 3-1 below gives the breakdown of the percentages of our power derived from each source. Table 3-2 shows us the emission levels when each type of fuel is burned in creating electrical energy. When this data is related to how much of these fuels we actually used in 2008 we obtain the weights of the indicated pollutants we pumped into the air in 2008. This data is provided in Table 3-3.

Table 3-1 Domestic Electricity Generation, Summary Statistics for 2008

Energy Source	Number of Generators	Net Generation (Thousands of Megawatt Hours)	Percentage of the Total
Coal	1,445	1,985,801	48.2%
Petroleum	3,768	46,243	1.1%
Natural Gas (and other gases)	5,569	894,688	21.7%
Nuclear	104	806,208	19.6%
Hydroelectric	3,996	254,831	6.2%
Wind	494	55,039	1.3%
Solar	89	864	0.0%
Biomass, (including wood and wood derivatives)	1765	55,034	1.3%
Misc. (incl. geothermal, pumped storage and other)	428	20,680	0.5%
Total	17,658	4,119,388	100.0%

It is important to notice the distribution percentages associated with each power source. Coal provides the largest percentage (over 48%) of our power. Petroleum provides only a modest 1.1% but notice also there are 3,768 generators creating this power. This may indicate that some of these power plants are not large and may be readily converted to natural gas which you see in Table 3-2 emits far fewer pollutants than coal. Natural gas is already our second highest fuel used for power generation at 21.7%. We have plenty of it so it is better choice than either coal or oil. Nuclear power at 19.6 % is our third largest source of power. Hydroelectric is the next largest at 6.2%. The rest of the power sources, wind, solar, biomass and a few others are all bit players at this time. But perhaps we can force changes there. We shall see.

The Pollution Situation Created When We Generate Electrical Power

There are two tables below. The first, Table 3-2, indentifies the pounds of various pollutants created in a million BTUs of energy input from the three major fuels: natural gas, oil and coal. The second, Table 3-3, identifies the actual tonnage of carbon dioxide, sulfur dioxide and nitrogen oxides that our mix of electricity generating plants shown in Table 3-1 actually discharged in 2008.

Table 3-2 Fossil Fuel Emission Levels (Pounds per Million BTU of Energy Input)

Pollutant	Natural Gas	Oil	Coal	Nuclear, Wind and Solar
Carbon Dioxide[9]	117,000	164,000	208,000	0
Carbon Monoxide	40	33	208	0
Nitrogen Oxides	92	448	457	0
Sulfur Dioxide	1	1,122	2,591	0
Particulates	7	84	2,744	0
Mercury	0.000	0.007	0.016	0.000

Table 3-3 Actual Emissions from Conventional US Power Plants and Combined Heat-Power Plants in 2008 (Thousands of Metric Tons)

Carbon Dioxide	2,477,213
Sulfur Dioxide	7,830
Nitrogen Oxides	3,330

The data in Tables 3-1, 3-2 and 3-2 were taken from DOE Energy Information Agency publications.

The Cost of Generating Electricity in the US

Costs Need to be Defined

Capital costs to build an electricity generation capability includes engineering, procurements and construction costs, plus the owner's costs for

9 Carbon dioxide is a pollutant only in the sense it is unwanted. At normal levels of concentration in the atmosphere it is not harmful. The other pollutants listed can be dangerous to our health.

land, cooling infrastructure, administration, associated buildings, project management, licenses and inflation and cost escalation. Construction costs include cost escalation and interest during construction. Financing costs are self evident. Operating costs includes operations and maintenance, fuel, and a return on equity.

Levelized costs are defined as the average cost of producing power (dollars per kilowatt hours or per megawatt) including all of the costs cited above over the plant's lifetime with provision for decommissioning and waste disposal. In essence it includes all of the costs divided by the lifetime expected power produced in kilowatt or megawatt hours.

Domestic Power Generating Costs

We look at cost calculations prepared by three different organizations. The differences in their results vary somewhat and are not unexpected. Differences in definitions of what constitutes a cost may vary a little as well as amortization periods for the costs of capitalized items and of course, the price of the fuel. Such things do result in differences. What the decision makers really needs to know when they have to select the fuel they want for a new power plant is what will be the likely cost per kilowatt hour when the plant is on-line and the power is sold to cost conscious customers. They therefore are interested in the rank order of output power costs to give visibility to the relative costs of one fuel versus another. By presenting results from three different sources we look to consistency to give us confidence in the rankings.

The first set of data is provided by the Nuclear Energy Institute in a paper called The Cost of Generating Capacity in Perspective and shown in Table 3-4.

Table 3-4 Levelized Power Costs (Cents per KWh)	
New nuclear capacity	8.34
Supercritical coal	8.65
Coal Integrated Gasification Combined Cycle (IGCC)	9.22
IGCC with carbon capture and storage	12.45
Gas combined cycle	7.60
Gas combined cycle with carbon capture and storage	10.31

In this set of data nuclear power is the least expensive, if you assume the carbon capture and storage will be a requirement. While Congress has not succeeded in its quest for a cap and trade bill, it may only be a matter of time. Even if it doesn't succeed, the EPA has declared that carbon dioxide can be regulated by the clean air laws and has plans to tax carbon dioxide emissions for power plants if a cap and trade doesn't get passed first.

Our second set of data comes from studies, one sponsored by California Regulatory Agencies and another by Lazard Ltd. Table 3-5 below provides relevant comparative cost of electrical generation from these studies.

Table 3-5 Comparative Cost Data
California Regulatory Agencies[10] and Lazard Ltd[11].

Fuel	Costs in Cents per Kilowatt-Hour (2008 Dollars)	
	California Regulatory Agencies	Lazard Ltd
Coal Supercritical	10.554	7.4
Coal Integrated Gasification Combined Cycle (IGCC)	11.481	10.4-13.4.
Coal IGCC with Carbon Capture and Storage (IGCC with CCS)	17.317	13.5
Gas Combined Cycle	9.382	7.3-10.0
Nuclear	15.316	9.8-12.6
Wind	8.910	4.4-9.1
Solar Photovoltaic (crystalline)	No Data	10.9-15.4
Solar Photovoltaic (thin film)	9.6	12.4
Concentrating Solar Thermal (CST)	12.653	9.0-14.5

A few words about the California Regulatory Agency data shown in the table indicates higher costs in California than does the Lazard data, which probably draws from the much larger national database of power generating

---

10 California Energy Commission and California Public Utilities Commission (May2008) by Energy and Environmental Economics Inc. Costs for the California studies are identified as "bus bar" costs and defined as prices leaving the plant including all capital, fuel and operating costs.

11 Lazard's levelized costs include production tax credits, investments tax credits, and accelerated asset depreciation as applicable. Assumptions are also made dealing with the amount of debt and the interest rate thereon. The California study undoubtedly made similar assumptions to arrive at their estimates but they were not listed. Most likely, differences in these assumptions are not consequential.

facilities. There is no question that California has realized severe air quality and earthquake problems that have influenced how their power stations are built and equipped. Higher capital costs are then expected in California and consequently the depreciation expense component of operating costs will be higher than otherwise.

Since the California Regulatory Agency data is unique and based on a relatively small sample (California for example has only four nuclear power plants) we regard that the Lazard data is more representative of US cost experience. The California data is presented here only because it is a very large state and knowledgeable people might inquire the reasons for an omission.

Making reasoned choices is our objective when we search for home grown energy to give us energy independence. In the final analysis then, we are interested in the relative costs of electricity and how each power source stacks-up on such factors as adjusted prices when pollutants are dealt with, reliability, fuel price stability, safety and perhaps other factors as well. To base our effort on more data than just the material above, we look at world wide experience in power generation.

Global Power Generation

The Organization for Economic Development and Cooperation (OECD) was set up in the aftermath of World War II to help rebuild the allied countries. It has survived until now and is the source of much useful data for us.

We show in the following presentation that in a global sense, nuclear electric power plants are fairly common. We'll look first at how pervasive they have become and then see how costs of power generation compare depending on the fuel technology employed.

Sixteen OECD countries have 344 nuclear reactors in operation with 13 more under construction (see Table 3-6). The electric utilities in the US have applied for licenses to build several more. It is relevant to note there have not been any significant accidents at any of these plants except our own Three Mile Island years ago. No one was injured in that accident. This is a truly good accident record.

Table 3-6 Nuclear Energy Generation in OECD Countries				
Country	Year 2008		As of October, 31 2009	
	Electrical Power Generated Terawatt-Hours	Percent of Total Electrical Power Generated	Plants Connected To the Grid	Plants Under Construction
Belgium	43.4	53.8	7	
Canada	87.9	14.5	20	
Czech Republic	25.0	32.4	6	
Finland	22.1	29.9	4	1
France	418.3	76.2	59	1
Germany	140.9	23.4	17	
Hungary	14.0	37.7	4	
Japan	240.5	24.9	54	2
Korea	144.0	36.7	20	6
Mexico	9.4	4.0	2	
Netherlands	4.0	3.8	1	
Slovak Republic	15.4	57.0	4	2
Spain	56.4	18.3	8	
Sweden	61.3	42.0	10	
Switzerland	26.1	39.0	5	
United Kingdom	47.7	13.2	19	
United States	806.2	19.6	104	1
OECD Total	**2162.6**	**21.5**	**344**	**13**

This country however, has not lost its skills in building such plants for the US market. Our companies have been involved in putting nuclear power plants on probably more Navy ships than we ever put in electrical power generating plants. If there is a significant difference in a nuclear power plant on a ship and one owned by a utility company it is only the size or power output. Otherwise they are very similar.

We generate 19.5% of our electricity using nuclear power. Eleven countries shown in Table 3-6 generate a larger share than the United States.

A Comparison of Power Generation Costs among OECD Countries

OECD or other international organizations have conducted a number of studies comparing the costs per kilowatt-hour to make electricity across the spectrum of countries having nuclear power plants. The costs are compared to coal, coal with carbon capture and storage, combined cycle gas turbines and wind onshore. Before laying out their study results we need to provide some background to define costs.

When comparing alternative long term investments, as with power plants the usual practice is to discount the stream of annual costs over the expected life of each alternative to create a "present value" for each. This permits alternatives which may have quite different streams of costs over different life-spans to be compared on an equitable basis. Economists may argue about what discount rate to use. To side step an argument the OECD people used both 5 and 10 % discount rates in developing the comparative data of electrical kilowatt-hour costs with five kinds of power plants in several countries. These data are provided below in Tables 3-7A and 3-7B.

Table 3-7A Electricity Generation Costs in OECD Countries
Cents per Kilowatt-Hour 5% Discount Rate

Country	Nuclear	Coal	Coal with CCS[12]	Gas CCGT[13]	On-Shore Wind
Belgium	6.1	8.2		9.0	9.6
Czech Republic	7.0	8.5-9.4	8.8-9.3	9.2	14.6
France	5.6				9.0
Germany	5.0	7.0-7.9	6.8-8.5	8.5	10.6
Hungary	8.2				
Japan	5.0	8.8		10.5	
Korea	2.9-3.3	6.6-6.8		9.1	
Netherlands	6.3	8.2		7.8	8.6
Slovakia	6.3	12.0			
Switzerland	5.5-7.8			9.4	16.3
USA	4.9	7.2-7.5	6.8	7.7	4.8
EPRI (USA)[14]	4.8	7.2		7.9	6.2

12 Coal with carbon capture and store means equipment installed to collect all of the carbon dioxide in the combustion products and store, probably deep in abandoned wells.

13 CCGT is combined cycle gas turbine that burns natural gas in a gas turbine and the residual heat is used to create steam for a steam turbine to make an efficient generator of electricity.

14 EPRI is the Electric Power Research Institute.

Table 3-7B Electricity Generation Costs in OECD Countries Cents per Kilowatt-Hour 10% Discount Rate					
Country	**Nuclear**	**Coal**	**Coal with CCS**	**Gas CCGT**	**On-Shore Wind**
Belgium	**10.9**	**10.0**		**9.3-9.9**	**13.6**
Czech Republic	11.5	11.4-13.3	13.6-14.1	10.4	21.9
France	9.2				12.2
Germany	8.3	8.7-9.4	9.5-11.0	9.3	14.3
Hungary	12.2				
Japan	7.6	10.7		12.0	
Korea	4.2-4.8	7.1-7.4		9.5	
Netherlands	10.5	10.0		8.2	12.2
Slovakia	9.8	14.2			
Switzerland	9.0-13.6				
USA	7.7	8.8-9.3	9.4	8.3	7.0
EPRI (USA)	7.3	8.8	10.2	9.4	15.5

The data in Tables 3-7A & B were derived from a sample of 170 power plants and are believed to reflect experience in 2008 or 2009.

Both the Lazard study and the OECD work show wind power costs spread over a wide range. The reason this occurs is understandable. The behavior of the wind is a regional phenomena blowing with different strengths at different times of the year. The Lazard data also has costs for power generated with solar technologies. Solar is also impacted by geography, particularly latitude, and local weather conditions so the costs of this too will show significant variability.

Electricity Price Stability

While nuclear power can deliver power at the lowest kilowatt-hour costs it has another distinct advantage when compared to natural gas. Its' price is relatively stable. Once built and operating, the price of fuel is a small part of the operating costs. Prices for nuclear power do not change as market prices for oil or gas fluctuate. Once the fuel is loaded into a reactor, it generally doesn't need refueling again for 18 to 24 months.

There is another piece of illuminating information that sheds light on why the operating costs are less for nuclear powered plants than for some other fuels. The DOE's Energy Information Agency in their 2007 summary of nuclear power statistics shows "Fuel Costs: nuclear vs. fossil steam 0.50 cents / kilowatt hour vs. 2.40 cents / kilowatt hour". The fuel for nuclear reactors is only 21% of the cost of fuel for fossil fueled generators. It is pointed out that the 2.4 cent price for fossil fuels used in the DOE computations is a number for a specific point in time. The prices of power derived fossil fuels, coal and natural gas, are sometimes volatile. Not so with nuclear reactor fuel

The Electric Utility Industry in the US

In the sections to follow we provide descriptions of the various ways electric power is generated, problems that characterize each power source and how we can deal with them. We previously indicated we will not deal with some power sources because of natural growth limitations.

While the electric utilities of the United States only derive about 1% of the power they generate by burning oil, our objectives are Energy Independence (minimal to no foreign oil purchases), and minimal emissions and later, complete elimination of emissions. As a big user of coal, the industry is a contributor to environmental pollution. Consequently we believe the US will be a better place to live if they convert to burning natural gas as an interim step or go directly to nuclear power or other non-polluting fuel.

Coal Fired Electric Utilities

Coal-fired power production provides the biggest percentage of our electricity as it does globally. It has had this position of prominence for many years, undoubtedly because coal is plentiful and inexpensive. In total, there are hundreds of coal-fired generating plants in the US. They cover a large range of sizes and ages. Some are almost a century old.

The EPA is doing its best to vilify coal fired utilities initially because of noxious emissions and lately because of carbon dioxide emissions. The regulations regarding control of these emissions are becoming increasingly expensive to implement thereby eliminating coal as a fuel. The fuel that has historically offered the lowest kilowatt hour costs now offers among the most expensive costs.

There also is doubt as to whether the older coal burning utilities can retrofit their facilities to capture and store carbon dioxide. Consequently, these issues might lead executives to conclude a conversion is the expedient thing to do. Some of the utilities are trying to modernize their coal burning utilities to remove pollutants and finding it is very expensive to do so. Duke Energy for example is currently (mid 2010) working to replace an old coal burning facility in Indiana with a 600 MGW plant that converts coal to gas from which mercury and sulfur are removed and the gas then used to power a turbo generator. The carbon dioxide created in this facility is to be captured and to be sequestered underground. Duke intended that this plant provide inexpensive power. Their plan may be shattered however as cost projections were re-estimated and have risen to $2.9 billion. The plant is still under construction with planned completion in 2012. Had Duke predicted what the ultimate cost of this plant might turn out to be, it might have opted for a nuclear power plant, possibly one of the new small modular reactors described later in this book. By installing 5 modules at 125 megawatts each, they could have had the 600 MW they need. [Duke's decision to build the complicated state-of-the-art coal plant with carbon capture and storage may have been the right one at the time, however, because the alternative small modular nuclear plants, discussed later, were not yet being offered].

We pointed out earlier that coal burning power plants contribute a great deal of pollutants that are detrimental to the environment. A lot of the pollutants have been known to be harmful to people's health but technology didn't offer adequate solutions. In 2002 a consulting company did a study entitled Particulate-Related Health Impacts of Emissions in 2001 from 41 major US Power Plants[15]. The study concluded that an estimated 4,800 to 5,600 premature deaths were associated with the 41 power plants being studied in 2001. It also estimated over 3000 hospital visits, 930,000 work days lost, and 111,000 asthma attacks from emissions. This study has been updated from time-to-time and other groups have made similar studies with similar results.

---

[15] Abt Associates November 2002 Particulate-Related Health Impacts of Emissions in 2001 from 42 major US Power Plants prepared for the Environmental Integrity Project, Rockefeller Family Fund, Washington DC.

Natural Gas Powered Utilities

Approximately 21.6% of our electricity was made using natural gas as the fuel in 2008. In that year we produced 20,561 billion cubic feet of natural gas and used 23,210 billion cubic feet. The short-fall was filled by importing about 4 billion cubic feet. About a billion cubic feet was exported. As a result of the development of horizontal drilling and the capability to fracture gas bearing shale rock domestic supplies are increasing dramatically implying a shrinking of imports, improving availability and gas surpluses that will exert downward pressure on prices. Since natural gas is much cleaner burning than coal it may be an ideal interim fuel for utilities whose long range planning includes nuclear power to convert from coal. If the government succeeds in imposing cap and trade or similar taxes the utilities may find it economical to think of gas as being a bridge fuel to use during the long design, licensing and building process associated with nuclear power. EPA provides the following comparison of emissions rates from coal and gas burning utilities:

> The average emissions rates in the United States from natural gas-fired generation are: 1135 lbs/ MWH of carbon dioxide, 0.1 lbs/ MWH of sulfur dioxide, and 1.7 lbs/ MWH of nitrogen oxides. Compared to the average air emissions from coal-fired generation, natural gas produces half as much carbon dioxide, less than a third as much nitrogen oxides, and one percent as much sulfur oxides at the power plant.

In Chapter 6 we provide natural gas resource availability data but that data does not reflect the potential from a massive increase in natural gas supplies that might be available when the technology to mine methane hydrates is developed. The methane hydrates exist in abundance in our coastal waters. DOE has been chartered by Congress to research availability of this energy supply and to develop techniques to bring them to the surface. [Methane hydrates are frozen crystals of natural gas and water existing under the pressure of at least 300 meters of ocean. To date we have not found a way to bring these to the surface without losing the natural gas.]

Conclusions Relative to Fossil Fuel Generated Power Plants

1. The predominant fuel for generating electricity is coal, but it is also very dirty. The technology to clean it of pollutants and to capture the carbon dioxide emissions adds to the costs and makes coal a very expensive fuel to use.
2. Only 1.1 % of our power is made by burning oil. We can make a contribution to energy independence, if we switched those plants to natural gas, and in doing so, we would also reduce carbon dioxide and other pollutants.
3. Natural gas and nuclear power are used in roughly equal amounts; natural gas is not free of emissions but it emits fewer pollutants than coal. Nuclear power is emission free but has great capital expense and long lead times associated with it. This indicates natural gas would be a good choice as an interim fuel for the several years that it will take to put additional nuclear plants on line. Ultimately natural gas fueled plants should be replaced with nuclear plants for reasons described later.

CHAPTER 4

Electrical Power Generating Plants

Fueled with Nuclear or Renewable Fuel Sources

THIS CHAPTER DEALS with nuclear, solar and wind as fuel sources. These power plants share a couple of important characteristics. Once built they provide relatively stable electricity prices and they do not normally emit anything to the environment The US has historically had low electricity prices which have facilitated the growth of our manufacturing economy. As pointed out earlier the prices of electricity created with wind and solar technologies, as they currently exist are fairly high. These price levels may be viewed by manufacturers as unfriendly and influence business decisions. There is nevertheless growing interest in wind and solar generated power and so it is a factor to be dealt with. Clearly given the interest in solar and wind efforts to find ways to reduce the cost of power so generated could be efforts well spent. We have more to say about solar and wind later in this chapter.

Nuclear Power

There are 104 nuclear power plants in the US. They are large installations (generally a gigawatt or larger) built at great cost and required years to build. They were so costly that financing risk became an issue as serious as the construction and other risks that the owners needed to contend with. These issues as well as an accident at the Three Mile Island installation resulted in the unwillingness of the utilities to build anymore such plants over the last several years. Both the issue of cost to build the plants and nuclear safety are addressed below:

A Very Short and Simplistic Primer on What the Reactor Does In a Normal Nuclear Power Plant

Nuclear fission is the process of splitting atoms. When an atom of the element uranium 235 is split by hitting it with neutrons, the nucleus of the atom results in two or more, lighter elements and free neutrons. The fission also creates kinetic energy and gamma radiation. Some portion of the released free neutrons are absorbed by other fissile uranium atoms causing further splitting of the atoms and creating a chain reaction. The kinetic energy converts to thermal energy that we can use in the same manner had the heat source been coal or natural gas or some other combustible product. Ultimately steam is created to drive a turboelectric generator to make electricity. The chain reaction can be controlled by using moderators characterized by their ability to absorb the free neutrons thereby throttling the chain reaction. Carbon is an excellent moderator as are water and some other materials. The moderators are also useful in converting the kinetic energy to the thermal energy that is most useful to us.

The control process may involve carbon rods that can be inserted into the reactor core to absorb the free neutrons. Absorbing more neutrons means fewer neutrons to perpetuate the chain reaction so inserting the control rods deeper results in reduced power output, and extracting the rods will increase power output. If the reactor is shutdown by fully inserting the control rods, there is still heat generated that persists for some time. The reactor design must accommodate this residual heat.

Nuclear Safety

Safety has been a much studied issue right from the beginning of the nuclear age. These studies resulted in a design philosophy in the US which was sharpened in the aftermath of the Three Mile Island mishap.

The fear in operation of any nuclear reactor is loss of cooling that could cause a meltdown of the reactor's core which, in turn, could cause severe damage and release of radioactivity. Radioactivity, of course, can be biologically damaging and potentially lethal. This was recognized early-on and the notion of building a containment vessel around all power reactors was adopted. Every power reactor built in this country has this protection. The containment worked in the case of Three Mile Island as the mishap was confined within the vessel. There was no damage from radiation or anything else outside of the containment vessel and there was no need to evacuate any of the plant's neighbors. There was also no radiation exposure to anybody within the facility. Part of the reactor did melt but has been isolated or repaired. Contrast this with the Russian Chernobyl incident. The Russian design criteria did not require a containment vessel. When a steam explosion occurred, the reactor melted down, killing initially 31 people. Radiation ultimately claimed more people with the final tally being 51. The plant was a total loss and is now buried under concrete. In due course, the Russian's learned much from us about reactor safety and have since improved their power-plant designs.

Much of the rest of the world were spectators to the early mishaps made in harnessing nuclear power and learned as we did. Now years later the only significant mishaps that have occurred among the 344 power reactors in the US and the OECD countries are the two just mentioned, Three Mile Island (1979) and Chernobyl (1986). So far the safety record is 14,000 cumulative reactor years with only these two major incidents[16]. Too bad this kind of record hasn't prevailed in regard to other means to fuel electrical power generating facilities. Coal burning electrical facilities, for example, are particularly distressing if you include coal mines accidents. Add the health impacts from coal burning power plants mentioned in the preceding section and one has to wonder if we would have ever burned coal had nuclear power been invented first

The World Nuclear Association has compiled global safety data covering the time span from 1970 through 1992 for all countries generating electricity. In their calculation they record fatalities including coal mine accidents (6400), hydropower dam failures (4000), and natural gas explosions (1200) all of which relate to providing fuel to run electrical generating facilities. They compare some very large numbers to the very few fatalities associated with nuclear reactor accidents (31). All of the reactor fatalities were associated with Russia Chernobyl accident. The 31 shown in their

[16] Recently Japan has experienced a nuclear mishap spawned by an earthquake and tsunami. This is discussed later in this chapter.

data were those involved in the immediate mishap and do not include the additional 20 people who eventually died of radiation poisoning. The data the World Nuclear Association dealt with ignored the long term impacts on the health of the population downwind of utilities burning coal.

Other factors contributing to the nuclear safety record include a design-in-depth approach involving reactor designs with multiple safety systems, quality design and construction, selection of reliable components and equipment, and redundant systems to control and mitigate damage to fuel rods and radioactive releases.

The main safety feature of most reactors is the designed-in negative temperature coefficient and negative void coefficient. This means beyond some design point, increasing temperature results in a reduction of nuclear activity, that is, it tends to turn itself off. The negative void coefficient means that if steam is formed in the cooling water there is a decrease in the moderating effect so fewer neutrons are able to cause fission and the reaction slows down automatically. This is a passive safety feature that doesn't require human intervention. [No forgetful or poorly trained operator can screw it up].

Radiation doses are controlled by use of remote controlled handling equipment for operations within the core of the reactor. Radiation dose levels are continually monitored as regards individuals and work spaces to ensure low levels of radiation.

The design of the reactor itself according to the Nuclear Regulatory Commission (NRC) requires a calculated probability of not more than one degraded core incident in 10,000 years. Overtime the industry itself has raised this to one event in 100,000 years and even longer in some of the best currently operating plants. In the aggregate, safety systems account for about a quarter of the capital cost of the reactor. Note the passive safety features were not present in some of the early reactor designs. Today the NRC will not license a utility to use a reactor that is not inherently safe.

The World Nuclear Association notes that if there is a weak point in the nuclear power plant system it is probably the operators themselves. Three Mile Island involved a minor equipment failure and failure of the operators to respond correctly. Operators are the weak point because the operation of nuclear power plants is inherently boring since they can percolate along for many hours or weeks without need for adjustment or tweaking of any sort. Improvements in operator training were required by the Nuclear Regulatory Commission after Three Mile Island and there have been no known problems since. It is noted the US Navy also has nuclear power on roughly as many warships as the US has stationary power plants. The Navy operates excellent training programs and it is reflected in the fact that they have never had a nuclear mishap of any kind on any of their ships.

Even though the reactors can be designed to be inherently safe and with very high reliability, having adequately trained power plant managers and operators is still vital. As we said, the Navy does an excellent job training people to run their shipboard power plants. So too does the Nuclear Regulatory Commission. This is another layer in the overall safety system to prevent mishaps.

What does all of this mean? Nuclear power is a lot safer than competing fossil fuels burned to generate electricity. And it won't pollute the air.

Having presented all of the above, we present the following impression of the Japanese Fukushima nuclear power plant damaged by the earthquake and tsunami in 2011.

Japanese Earthquake/Tsunami/Nuclear Reactor Catastrophe

Japan has suffered a terrible catastrophe involving their nuclear power industry. It has been suggested by some people that we should close down our nuclear power plants, all 104 of them. Wiser people realize that we shouldn't close down anything until we find out what happened and then try to make such improvements that are appropriate so it doesn't happen to us. While we don't know all of the facts yet, we have some information that is useful. The facts as we know them one week after the event: The nuclear reactors were built on a known geologic fault line with design features to minimize earthquake damage. They were also built on the shore of the Pacific Ocean to give them access to the ocean for cooling purposes. They anticipated earthquakes but they thought that an earthquake in conjunction with a tsunami to be highly unlikely. That was a mistake. As events unfolded the earthquake did little damage to the nuclear plant in spite of the severity (8.9 on the Richter scale). It was followed within an hour or so by a 46 foot high tidal wave that did major damage[17]. It knocked out the electric power that supplied the primary coolant pumps on all of the reactors. The reactors at that point turned themselves off. The nature of this type of nuclear reactor, however, is that they continue generating heat for several days after being turned-off. The amount of heat is predictable as it follows a known "decay curve". Initially a lot of heat needs to be removed but gradually the heat output declines to very little. This heat needs to be removed because it can damage and ultimately destroy the reactor. The backup cooling system consisted of diesel electric generators which powered the coolant pumps. The tsunami destroyed this backup system and the real trouble began. The designers of this system years ago never anticipated an earthquake triggering an almost immediate tsunami. The specific geologic fault on which these reactors are situated contains a subduction zone where two tectonic plates overlap. This can lead to an earthquake that can be accompanied by a tsunami. Perhaps the state of understanding of the seismic data was not so well developed years ago when planning for these power plants was done. Now months after the incident it is known that there were many fatalities from the earthquake and the tsunami. At least two employees of Tokyo Electric were killed at the site of the power plant. Others will undoubtedly succumb from radiation exposure accumulated over time. Some time needs to pass to see how this evolves[18]

The US Nuclear Regulatory Commission Responded to the Fukushima Incident

The US Nuclear Regulatory Commission responded to the Fukushima incident by establishing a task force to review the Japanese experience and make recommendations to minimize the possibilities of a similar

---

17 The nuclear facilities were protected by a 30 foot high wall.

18 An article published by Fox news on July 23, 2011 entitled Nuclear Power Is Extremely Safe—That's the Truth About What We Learned From Japan by Alex Epstein of the Ayn Rand Institute indicated there were no deaths and no personnel experienced radiation exposure beyond safe limits resulting from the reactor damage in the wake of the earthquake and tsunami. The two deaths indicated in the text above apparently resulted from the tsunami or the earthquake.

incident happening in our nuclear power industry. The Task Force Report recommended the following immediate actions[19]:

- "Order nuclear plants to re-evaluate seismic and flooding risks. Update structures and components to provide more protection.
- Order plants to have back-up electricity for spent fuel rods and better monitoring and cooling.
- Update emergency plans to incorporate severe-accident guidelines that now are mandatory.
- Order plants like those in Japan to have better ways to vent dangerous hydrogen gas and prevent explosions.
- Order plants to prepare for disasters that damage more than one reactor."

The Report also recommended these longer-term actions:

- "Require plants to re-evaluate seismic and flooding data every 10 years.
- Require plants to have systems in place to ride-out extensive blackouts.
- Improve emergency training for plant operators."

An Expert's View of Nuclear Power in the Wake of the Japanese Nuclear Power Disaster in 2011

William Tucker is the author of a book entitled Terrestrial Energy: How Nuclear Power Will Lead Green Revolution and End America's Energy Odyssey. He writes in the Wall Street Journal[20] to summarize his reaction based on his knowledge of the electric power industry. In the WSJ article, he notes that other methods of generating electricity are not without risks. For example he indicates a natural gas powered electrical generating station in 1944 blew up killing 130 people. Coal mining accidents killed 100,000 people in the 20th century and still kill an average of six people every day in China. A hydroelectric dam collapsed in Japan during a recent

[19] Wall Street Journal Article July 2011, Nuclear Plant Safety Proposals Put on Fast Track.

[20] Wall Street Journal. April 23rd 2011. Why I Still Support Nuclear Power, Even After Fukushima

earthquake washing away 1800 homes with an as yet uncounted loss of lives. He goes on to ask why we have allowed this to happen. He responds to his own question with the assertion we have no better alternatives. Consider solar and wind and hydropower and a few relevant facts. "The amount of energy harnessed by wind and water is about 15 orders of magnitude less than what can be derived from uranium". He indicates as an example that the Hoover dam requires a 250 square mile lake to generate as much power as a nuclear reactor sited on one square mile. Wind mills also require a lot of space. Replacing a one megawatt reactor would require many 45 story high wind mills stretched a quarter mile apart the length of the distance between New York and Albany to generate as much power and, of course, they would do so only when the wind blows. Solar collectors to generate as much power would cover 20 square miles of photovoltaic cells or polished reflecting mirrors to focus the sun's energy. And it would only work when the sun shines. It was pointed out in an earlier chapter of this book that because wind and solar are not steady performers the energy produced cannot be relied upon for base power; it is useful only to help satisfy peak power requirements. As a consequence of these factors, nuclear power becomes far more attractive, particularly in light of the fact that technology has moved and continues to move in the direction of safer nuclear systems. Remember the Fukushima reactors were over 40 years old.

Important New Technology Developments in Nuclear Power Plants are Underway

In 1979 the US had a nuclear accident at Three Mile Island that resulted in a partial meltdown of a reactor. The backup cooling system failed. In the post mortem, the finding was this was caused by human error. As mentioned earlier, no one was hurt and there was no radiation leakage.

The engineers who design reactor systems, nevertheless, decided they could minimize possibilities of human error by creating passive backup cooling systems for future boiling water reactors. Unfortunately the Japanese reactors were too old to benefit from this safety improvement. Not so, however, relative to the reactors that will be installed today. The passive backup systems automatically trigger the backup cooling system with no human intervention and provide a few days' supply of water to absorb the decay heat without an auxiliary electrical supply.

Table 3-8 lists new nuclear reactors being considered by the electric power companies plus a few that are under development. The latter machines are unlikely candidates for power company consideration because of the time required to make technology advances in this field as well as the long time required by the Nuclear Regulatory Commission (NRC) to certify a design as acceptable. Only one, the Westinghouse Electric AP 1000 has been certified, so far, and it is a large 1117 MW reactor akin to the very large reactors that were installed in the US years ago and remain in operations today. These systems were very expensive and with inflation overtime have become even more so. The AP1000 is an improvement in a technology sense compared to older machines. Among other improvements, it contains a passive back up cooling system. The huge expense associated with the large power plants introduced financing risk which the electric utility companies when selecting a power plant design. Hopefully the companies that selected the AP 1000 have the financing under control.

Companies contemplating new nuclear plants now have an interesting alternative: Small Modular Reactors, or SMRs, which are being designed by several companies to overcome financing and cost problems experienced in the past. These new machines can be installed in groups or individually as needed or when funding is available.

Table 4-1 Large Reactors

Vendor	Product Description	License Application for certification to be filed with NRC	Advantages	Disadvantages
Westinghouse Electric (Long history building nuclear power plants)	AP1000 PWR (pressurized water reactor); Fuel produces heat creating steam to drive turbo-generator. Generates 1117 MW of electricity. (PWR reactors are more often used than other types).	Applications filed and approved for 4 in China and 6 (3 utilities) in the US. A final approval is pending and holding up construction start-up of US plants.	New passive safety features rely on air flow, pressure changes and gravity rather than electrically driven pumps or people. No safety features needed other than the passive cooling. Water cooled.	Needs refueling every 18 to 24 months. Cooling water is at high temperature and pressure requiring periodic maintenance. Lots of water needed.

Table 4-1 Large Reactors				
Vendor	**Product Description**	**License Application for certification to be filed with NRC**	**Advantages**	**Disadvantages**
Small Modular Reactors				
Babcock & Wilcox (Very experienced)	mPower. 125 MW electrical output modules. As many as 10 modules can be Interconnected.	This company may be closer to filing for certification than the others. First plant installed by 2020 if regulatory approval is timely.	Modular construction. Modules can be added without disrupting operations. Financing risks minimized. Built largely underground to facilitate containment. Partnership with Bechtel. Subassemblies can be rail transported. Spent rods can be stored underground on site. mPower includes reactor and steam / generator for an integrated solution.	B&W doesn't admit to any.
NuScale Power (New company)	45 MW PWR for 1 module. Full scale plant to have 12 to 24 modules for 540 to 1080 MW of electricity. Water cooled and moderated.	Plans to apply for design certification in 2012.	Passive design. Water cooled but no pumps used, only convection. Refueling a module can be done without impacting other modules. Modules largely manufactured off-site. Fuel and steam generator housed in water submerged steel vessel with greater ability to withstand pressure and dissipate heat than conventional building a traditional PWR	Refuel every 24 months. To refuel reactor would have to be removed from the containment vessel and transported to a service area where refueling is done with remotely operated machines.

Table 4-1 Large Reactors				
Vendor	**Product Description**	**License Application for certification to be filed with NRC**	**Advantages**	**Disadvantages**
Hyperion Power Generation (New company)	Liquid metal cooled reactor 25MW electricity (Liquid lead bismuth). HPM power module is a "Fast" reactor. (Fast reactors do not require a moderator as do PWR reactors. Fast reactors use large mass of fuel to provide many neutrons	Certification in the US will take 3 to 5 years once application is filed. (No info on when filing is expected).	No refueling required; entire unit replaced after 8 to 10 years. (This may be a disadvantage depending on cost.). Lead bismuth coolant safer than sodium cooled reactors since leak in sodium reactor which exposes the sodium to water is dangerous. Lead bismuth leak would seal itself when the metal solidifies.	Hyperion will "take care" of disposed reactor. There is little experience in using fast reactors in the electric power field as most of them have been used for research or military purposes.
Toshiba (Big company)	Toshiba 4 S (Super, Safe, Small and Simple) 10MW liquid sodium cooled fast reactor. Sealed reactor never to be refueled.	Plan to file for design review with NRC in late 2012.	30 year life, then entire reactor returned to Toshiba to swap for another. Can sit mostly unattended for 30 years. Fast reactor can fission some of the longer lived isotopes in the spent fuel reducing the volume of waste. Plan is to bury the reactor underground.	Toshiba will "take care" of disposed reactor. Waste is highly radioactive. Design description presents a unique device which may be very experimental.
Run-off between General Atomic and Westinghouse for a DOE grant	Next Gen Nuclear Plant High Temperature Gas Cooled Reactor; 112 to 270 MW electricity.	8 to 10 years to design once DOE selects the contractor	Design expected to operate at high temperatures (900 degrees C) where the heat could be directed to certain industrial purposes.	This is not an option for anyone wanting a nuclear power source over the next several years. Focus is just on the reactor and not an integrated power plant to generate electricity.

Recycling of Spent Nuclear Fuel Rods

Fuel rods in nuclear power plants need to be replaced when the fuel is "spent". There are basically two ways to deal with these highly radioactive spent rods: (1) recycle them or (2) bury them to hide them forever so they will expose no one to radiation: Forever in this case means thousands of years. The USA ruled out recycling when Ford was president because of the possibility that plutonium in the spent rods might be stolen and used to build bombs. The US sought to bury the rods at Yucca Mountain in Nevada but it never happened. Not in my backyard (NIMBY) problems. The Government recently decided that the rods will never be buried in Nevada; however, some members of Congress are trying to force reassessment of this decision.

Every country in the World that has a nuclear reactor, except the USA, has their spent rods recycled[21]. No plutonium has ever been stolen by anyone anywhere. There are several facilities where fuel rod recycling can be done in such places as France, Great Britain, Russia, and Japan. A paper from the World Nuclear Association updated in March 2010 indicates that "so far 90,000 tonnes" of spent fuel rods from commercial power plants have been reprocessed.[22] Incidentally, no one has ever been as foolish as to attempt to steal plutonium since it is highly radioactive. A thief would be inviting more trouble than anyone wants.

While the US has never recycled the spent fuel rods from our utilities we nevertheless have the capability to do so. The Department of Energy Savannah River Site near Aiken, South Carolina has been supporting the country's nuclear program for years and in many ways. Currently it is building a processing plant to convert plutonium in excess nuclear warheads to a "mixed oxide" (MOX) fuel for certain nuclear reactors. This process is similar to recycling reactor spent fuel rods. We believe the State of South Carolina should welcome the business of recycling the spent rods. To them

---

21 It appear the spent fuel rods at Japan's Fukushima nuclear reactor site were stored on site and contributed to their problems following the earthquake and tsunami. Japan does recycle their spent fuel rods but these must have waiting to be shipped to the recycling site.

22 World Nuclear Association, <u>Processing of Used Nuclear</u> Fuel updated 3/2010

it is industrial development with an opportunity to create high paying jobs in a State that needs it badly. There would be no possibility of plutonium recovered by recycling being stolen from the Savannah site. The recovered plutonium isolated can be directed to the MOX facility mentioned above. In essence, all the plutonium is converted to reactor fuel.

Aside from solving the problem of what to do with spent fuel rods there are benefits to be derived by recycling:

- Recycling involves elimination of unwanted spent material, saving unspent fuel that remains in the spent rods and addition of new fissionable material to create new fuel rods. The process saves roughly 25% of the energy that was originally in the rod and thereby extends the use of uranium. This enhances energy security because the raw materials (uranium ore) used to make fuel rods come from foreign sources.
- Perhaps more importantly the process also converts some of the highly radioactive spent rod material into radioactive waste of much shorter half-life that after about 100 years will have totally decayed. A volume of radioactive material reduced in size to about a fifth of the volume before recycling will still require long term storage in a place like Yucca Mountain. This means that if Yucca Mountain is ultimately used it will have available capacity to last us quite a while.

The economics associated with spent fuel reprocessing are uncertain because of several situations including different physical, chemical and nuclear compositions of the rods from different reactor designs and the changing nature of the recycling processes resulting from improvement efforts. The economics are also clouded by the fact that the federal government years ago levied a tax on nuclear utilities. The money was to be used to solve the problem of what to do with spent fuel. They used billions to create the repository in Nevada that they now don't want. As it happens, the spent fuel rods now stored on the utility's sites have accumulated to a volume that comes close to filling the underground repository. This means even if the decision is made to store the rods in Nevada we will still have to address the issue of what to do next? Expand the storage site, find another site or recycle or some combination of these? Since we spent many years to create the hole in the ground nobody wants in their backyard, it seems foolish to start down that path again. Our only recourse is to recycle the spent fuel rods. The most logical place to do this is the DOE's Savannah River Site in South Carolina for the reason stated above.

Conclusions Relative to Nuclear Power Plants

1. The nuclear power plants that have been built in the US are all large (average over one megawatts), took a long time to build and are expensive, so financing is a significant risk.
2. Reactor manufacturers have new designs in process for small modular reactors. These are typically being designed with these objectives in mind: increased affordability because they can be bought and installed in relatively small modules and added to when the need arises, factory built to achieve economies, shortened build schedules to reduce costs, and mitigation of financing risks.
3. Worldwide statistics based on experience with over 400 nuclear power plants spanning a fifty year period support the fact that nuclear electric power production is safe. There have been three accidents each of which has contributed to our knowledge base: Chernobyl [which taught the Russians that they must include a containment vessel], Three Mile Island [which taught us (a) that humans are fallible and must be better trained and (b) passive back up cooling should be required] and more recently Fukushima. [Which taught us a lot about locating nuclear power plants in earthquake prone areas over a tectonic subduction zone to the ocean] While the earthquake and tsunami caused immense loss of life, there were no casualties from the damage done by the incident to the power plants. Contrast this with the casualties indentified in the text associated with the use of coal and natural gas over the last century or so. Note also that the only accident involving a US nuclear plant was Three Mile Island in which no one was lost or hurt. The Nuclear Regulatory Commission never-the-less learned from this experience as well as the recent Fukushima incident and imposed new requirements on reactor and power plant designers. I conclude they have done a good job of ensuring that our use of nuclear power is safe.
4. Nuclear Power delivers abundant energy at the lowest kilowatt hour costs without polluting the air.

Electricity Generation With Solar and Wind Power

Although the total field of renewable energy includes many sources and technologies, including hydro, ocean, biomass, the utilization of solar energy and wind energy has tended to be the dominant topics of interest. The interest in these areas has been appreciable and gains most of the attention of government, commercial and related advocacy groups. Solar electric power and wind powered systems are the principal focus. Within solar electric power, moreover, photovoltaic arrays and solar thermal plants are the main technologies of interest. This discussion of solar energy is constrained to solar power plants of sufficient size to deliver power to the electric grid system, of which photovoltaic arrays and solar thermal collectors are the two significant topics.

Solar and wind power share a few common characteristics. They are both pollution free and do not require any fuel other than what Mother Nature provides every day. They are also both highly intermittent. Solar electricity generators works only when the sun shines and they work best where the sun's intensity is strong. Lower latitudes are better than northern locations. Arizona is a great place for photovoltaic solar panels. New England is by contrast a poor place because of its' northern latitude and frequent cloud cover. Generally any locations such as the north west coastal area and northern states bordering the Great lakes have too much cloud cover for too much of the year to support large scale solar powered electricity generation. Thus solar power generation works well in some areas, not so well in others. Wind also has to be considered a regional resource. Some places receive considerable wind and some do not. It will work best where the wind blows a lot and fairly strongly. Offshore wind farms are being considered since they sometimes offer better wind intensity and duration.

Since the electrical energy created by solar and wind is very intermittent it is not the kind of power the nearby utility company can rely upon 24 hours a day. If the energy generated by wind or solar were timed to coincide with peak power demand, it could fulfill a need for the utility company. A bigger problem is the electrical grid discussed below. If the solar generating capability or wind farm is too far from where the power is needed, transmission losses on the grid may be too great since much of the power is lost. If we had large batteries to store power made by wind or solar collectors until the utilities need it, these power sources could be better accommodated. But batteries of the required scale have yet to be developed.

As currently configured, the grid is delicately balanced between supply and demand for electricity. A surge in either can upset its' balance and it can fail. The unsteady nature of the power generated by solar or wind was at one time a problem for utilities because it aggravated grid control problems but sophisticated power management systems have been developed which overcome this problem.

In summary we have identified some limitations of commercial grade wind and solar power plants: They have to be carefully designed for the specific geography and weather conditions, and be consistent with the needs of the utility and its' power transmission system. Note also in a an earlier chapter that the cost of wind power is high. The fuel may be free but the required capital investment and maintenance drive up the KWH costs, particularly for offshore installations. Given the problems mentioned here with solar and wind electricity generators; we really don't know the extent to which these systems can alleviate our need for clean electricity generating capacity. But the readers must be realistic about what can be accomplished in this aspect of power generation. Recall in Figure 3-1 we showed the amounts of electricity generated in the US in 2008 from the available fuels. Wind provided only1.3 % of the amount generated. Solar, in which 89 installations provided 864 megawatts of electric power, did not even register. This means it did not provide even one tenth of one percent of the power we needed. More recent data relative to solar energy generation were checked. The data changed slightly. With continuing research and development, the percentage contribution from solar energy may change in time but don't expect miracles. Consequently the message is solar cannot be expected to generate meaningful amounts of power. Wind energy will grow but is unlikely to generate sufficient power to become a significant factor in meeting our total need for electrical power. What this country needs is lots of cheap electricity to run a growing economy and facilitate manufacturing industries to provide jobs. If we are to avoid significant impact on our environment, nuclear power is our only option.

Conclusions Relative to Renewable Fueled Power Plants

1. Other emission free fuels are solar, wind and hydropower. Hydropower is not a significant growth contender because there are no major rivers that could be dammed that have not already been dammed to provide the falling water power source. Solar and wind are a source but they have limited potential to deliver inexpensive reliable power. Locating solar and wind electrical generating plants is complicated. Nobody really knows what the potential is for these power sources because finding suitable locations may severely constrain the possibilities.
2. The other renewable fuel sources on our list are biomass, geothermal, and pumped storage[23] and unlikely to grow into major contributors.
3. Nuclear energy, as indicated above, is our only option for meeting our total energy requirements at the lowest cost per kilowatt hour while emitting no unwanted and potentially damaging emissions.

---

23 Pumped storage refers to a utility company that has a high reservoir into which water can be pumped when it makes more electricity than customers want. Later, when power is needed, the water is drained through turbo-generators to increase electricity production.

CHAPTER 5

Power Generators Need a Grid to Distribute Electricity

The Grid Has Problems That Need to Be Addressed

THE GRID INCLUDES the familiar network of tall steel towers that support the cables that carry electricity from the generating facilities to their customers. These cables crisscross the landscape. Although unattractive, they are passive, make no noise and emit no foul odors. We generally ignore them. They can be a problem, however, because grid failure is responsible for the occasional black-outs that darken cities and trap people in elevators. A grid failure somewhere can cause a loss of electrical power over large areas because control of the grid is fairly primitive allowing failures to propagate rapidly. A grid failure that results in blackouts causes economic problems to people and the companies that are impacted. Perhaps a larger problem is grid power loss. Energy loss in the cables and transformers of the grid occur throughout and are dissipated as heat to the atmosphere. According to the International Electrotechnical Commission (IEC), the leading global organization that publishes international standards for electrical, electronic and related technologies, grid losses between power plant and consumers can easily aggregate to 8 to 15%. Considering the amount of power generated, these losses are significant.

The power companies get paid for the electrical energy consumption as read on their customer's meters. The energy lost on the grid never reaches a meter so the electric utility must absorb the costs which it can't afford to do. It therefore builds the costs into its rate structure. More importantly, it must generate power that will ultimately be lost. If it is a coal or natural gas burning utility, it will be generating electricity but also emissions that nobody wants. These too must be dealt with using appropriate capture systems. This could be a significant expense. Consequently investment in the grid to improve its' efficiency by reducing line losses is money well spent. Such investment could be viewed as conservation that avoids using fuel to make electricity that would otherwise be lost.

Background of the Grid

A few decades ago each utility built its own grid to distribute power to its customers. It was a local affair. The utilities were then heavily regulated industries that were able to pass costs, including the grid related costs, to customers. Sometimes the local utility's grid was connected to another utility company's grid so they could buy or sell power to each other when a need arose. Ultimately a grid crisscrossing the country was created. Since it had never been designed to be a national grid, it had deficiencies that could have been avoided had it been planned as an integrated design from the outset. For example, it was not designed to transmit large amounts of power across long distance nor was it designed to quickly isolate failures to prevent them from propagating over large areas. It would be foolish to modernize our electric utilities to efficiently make low cost electricity without also modernizing the means to distribute the power they create.

At some point the government deregulated the electric power industry and the companies began to evolve into competitive businesses. New owners bought into the industry or created new power plants with little investment in distribution facilities. Some companies concentrating on their profitability sold their grid holdings to others. Power plants with excess capacity sometimes sold power to very remote customers. California is a case in point. Since it has earthquake and air quality problems they would rather buy power than create more power plants. Thus, the need for long distance transmission became a necessity and imposed loads on the grid that were not originally anticipated. The profit orientation of the newly de-regulated utilities resulted in neglect of grid maintenance ultimately leading to fragility and occasional breakdowns and black-outs. Summing up; the grid became the fragmented stepchild of the electric power industry with no single organization in position to step in, take charge, assess its long term needs and provide the financial resources to make major improvements to accommodate future system loading.

The DOE's EIA reports the following data that puts dimensions to the issues involving the grid. The Federal Energy Regulatory Commission (FERC) has approved more than 850 applications to sell power competitively on the grid. The grid is a huge network consisting of about 211,000 miles of high voltage cables.

Expected Requirements of a Twenty First Century Grid

Think for a moment what will happen to the load on the grid as new additions of automobiles to the inventory become hybrids and particularly plug-in hybrids and later all electric cars and trucks. 220,000,000 plus cars and trucks fueled by the electric power industry. Can the grid handle the load? Not a chance! How about normal population growth? It goes on inexorably. Big implications for the grid! We must be ready with a grid of significantly larger capacity that can handle large loads with minimal transmission losses.

People who understand such things have studied the grid and come to conclusions about what has to be done. One source has defined the upgrade requirements as follows:

An Extra High Voltage Backbone Is Needed

A backbone for the national grid that would allow the utilities to sell their power is needed. It requires very high voltage to minimize line losses. The capability to move power over long distances with minimal loses will permit competition among utilities and result in lower costs for consumers. Lower power costs will also facilitate economic development, particularly in manufacturing where low power costs are essential. We don't have a true backbone on the grid today because of the haphazard way the grid was created. There are losses all along the length of the wire that are higher in stretches operating at low voltages. Sections with higher voltages sustain lower losses. These losses can be significant but are controllable by rebuilding the low voltage sections to run at higher voltages. Today portions of the grid operate at voltages up to 765,000. A restructured grid will have a high capacity backbone operating at the high voltage that will interconnect major power producers to their remote consumers.

Note: superconducting electrical cable has been invented that has no line losses which saves money, but isn't cost competitive with conventional cable. This type of cable needs to be under ground to facilitate holding the cable at the temperature of liquid nitrogen. This is –321 °F. Any warmer and the nitrogen will boil off into the air. While it has virtually no line loss, it is expensive and has an operating cost keeping it at the low temperature required to maintain super conductivity. Someday, hopefully soon, the relevant technology will advance allowing a superconducting solution for the grid to prevail in cost-benefit studies. The day may be fast approaching. American Superconductor recently sold a ten mile stretch of superconducting cable in South Korea.

Enable Energy Efficiency by Converting to a Digitally Controlled Grid

At this time the bigger problem is creating control of the grid and a program which will restructure the grid to meet the future load demands and reliability that it should have. This problem has not been solved but is well understood by experts in the industry and steps have been taken by the federal government and others to attack the issue. A key part of the digitally controlled grid is being addressed under the generic title of "Smart Grid". Smart Grid is focused on giving utility companies real time information about their customers power needs so that it can better match output to their needs. The key technologies addressed are integrated communications, sensing and measurements, advanced components, advanced controls and decision support. Integrated communications can be provided by any of the well developed technologies that are available. It is essential to know and anticipate the demands for electricity throughout the utility's domain. Sensing and measurement instrumentation will allow data to be analyzed and converted into information to support decision making and control of the system. Advanced components technology includes such things as "smart meters". These devices will keep track of power usage on a time-of-day basis and will give the utility the opportunity to have differential billing. This will permit it to offer cheaper rates for night time usage to induce people to shift discretionary electricity usage to evenings. This is when the power company usually sees a lighter load than daytime usage. Load leveling is accomplished and relieves their peak load demand. Power companies have to meet peak load demand. By dropping the peak load they may be able to defer capital expenditures to augment their generating capabilities. This is going to be very important to both the power companies and consumers when "plug in hybrid" cars become available and people will want to recharge their cars overnight in their garages.

The power company will also want to monitor some appliances for customer's power demand on a 24 hour basis. All this data is processed and displayed by the power company to give it the tools to better manage its business.

Real Time Transmission Monitoring

The electric utility industry has always been faced with matching the demand for electricity with the output of its generating and distribution systems. Two areas have been problematic: common coal burning boilers and nuclear reactors because by their nature their output cannot be changed quickly. Their whole purpose is to provide the heat to boil water and make steam that in turn expands passing through turbo generators which cannot be quickly cranked up or turned off if the utility needs more or less electricity. The utility companies can deal with this if they can anticipate it. Natural gas fueled systems burn gas which drive turbo generating equipment are much more responsive to changes in demand. The power companies, not really knowing much about their customer's needs, change their power output based on historical time-of-day and date data to match expected need. They do a good job at it but are occasionally flummoxed by breakdowns in the distribution system. Isolating the breakdowns take time. In the meantime, the system can becomes unstable because demand and the utilities output are mismatched and the system can fail. What the utility companies need, to avoid this type of problem is real time monitoring of the entire system, so it knows immediately when and where problems occur. This is mainly a technology issue that is being addressed.

A Suggestion for the Utility Industry: Take Control and Build a Modern Grid

Many people in the industry and in certain government agencies have recognized the problems with the grid, but their efforts to solve the problems are only beginning. The on-going efforts focus on design and reliability standards, efficiency improvements and specifications for the modernized grid. The grid is owned by the utilities that built it and use it, not the government. The Federal Energy Regulatory Commission (FERC) in 1999 issued Order 2000 calling for the electric utilities to form regional transmission organizations (RTOs) that will operate, control and possibly own the power transmission system. They undoubtedly did this but it doesn't seem like they accomplished much in the past 10 years. They are now identified as Regional Reliability Organizations. They work with the company doing reliability and standards work identified as the North American Electric Reliability Corporation (NERC). While they promulgate reliability standards they don't own anything. So a consortium of the major utilities need to get together and create a mutually acceptable plan to jointly own the grid, raise the financing to update the grid structure and make it into a 21st century grid. They can call themselves the New Grid Company. Since the utility companies generally have excellent credit, a consortium of them guaranteeing bonds should have little trouble finding the necessary funding. Companies using the new grid should pay for usage and the proceeds can pay the operating costs of managing the New Grid Company and repaying the bonds. Someday the New Grid Company may pay dividends to its owners. All it takes is leadership to get it going.

Conclusions Relating to Electric Power Generation Grid

The transmission grid is inadequate and needs to be overhauled and expanded. The problem is being studied and ultimately a specification for a 21st century grid will evolve. A big problem remains: Who will emerge as the leader to undertake the construction management, and the operation of the project, and work out the financing?

CHAPTER 6

Domestic Availability of Energy Resources

Introduction

THIS CHAPTER ADDRESSES the availability of oil and natural gas in US territory for the purposes of showing that there is no need to go to other countries to buy the 11,000,000 barrels of oil we buy every day. While this chapter addresses only the issue of availability, the next chapter addresses the impact on the US economy of exploration and production to obtain these resources and reduce the imports we can no longer afford.

We have on land and in our coastal waters potential reserves that would last many years as we move forward in developing more efficient transportation fuels and clean energy sources for power generation. Part of these reserves result from improved technologies for locating and drilling on existing resource bearing properties. Much of these reserves, however, are in areas blocked by the government's moratoria on drilling on Federal lands and off-shore areas. In this chapter we illuminate the information that is available and how it was obtained and the assessments by experts that indicate the extent of reserves that might be tapped if the moratoria were lifted. You will see that we probably have far more oil and natural gas resources than anybody might have imagined only a short while ago. But first, we look at the current situation with natural gas, since it is being exploited in many places resulting in an abundance of this fuel.

Overview of the Domestic Natural Gas Situation

Compared to oil, we import relatively little natural gas. Table 6-1 below provides statistics on natural gas production, consumption and imports.[24] Using the data in the table we calculate imports being only 15% of consumption. This calculation does not include the liquid natural gas (LNG). New drilling technology will result in more gas being found in the US, resulting in a declining need for natural gas imports.

Table 6-1 Highlights of the US Natural Gas Industry—2010 Data	
U.S. Production (Dry gas production)	21,577 billion cubic feet
U.S. Consumption	24,133 billion cubic feet
U.S. Imports	3,737 billion cubic feet
Imports as a Percent of Consumption	15

New Technology Improves Drilling Efficiency and Minimizes Environmental Impacts

Historically, the petroleum and natural gas drillers have been unable to extract more than a modest percentage of the product from their wells. In recent years, however, the industry has developed new technology to increase the amount of oil or gas that can be obtained. This technology has permitted drillers to reopen old wells to obtain additional oil or gas. Two developments have given us this capability: horizontal drilling and the technology to fracture shale rock at points along the horizontal wells that contain oil or gas. Together, these technologies have permitted significant improvements resulting in more productive drilling and importantly a cleaner drilling situation. This means fewer well heads on the landscape and in case of accidents, fewer possibilities to soil the environment. The second technology, fracturing shale involves sending water and chemicals at high pressure along the horizontal wells to break the shale to release oil

[24] Department of Energy, Energy information Agency, Natural Gas Summary

and gas. Scientific American offers the following explanation of the process included in the footnote.[25]

The process has been used for a while and some problems have become evident, particularly because of the chemicals used. For the most part, however, few problems have occurred because the fracturing often occurs deep underground and far below the water table and the aquifers where some communities obtain their water supplies. Congress has noted this and asked EPA to determine if new regulations may be required to minimize problems. This is a work in process.

There are vast areas in the US having gas bearing shale leading some to predict the US now has natural gas to sustain us for many years. This has already resulted in a drop in natural gas prices to around $4.00 per thousand cubic foot. Since we are now finding large quantities of domestic natural gas it is unlikely natural gas prices will rise in the foreseeable future, so use of natural gas for transportation and power plant purposes is now a viable and relatively safe alternative to imported oil. Note that the price of imported oil will increase as we extricate ourselves from protracted recession. Consequently, the use of domestic energy resources will become increasingly attractive.

There is one other situation that may impact the future supply and uses of natural gas: methane hydrates are present in coastal waters off shore north of Alaska, and the east and west coasts of the continental US. This is discussed briefly below.

Methane Hydrates

Natural gas is mainly methane. It has been known for some time that there is an abundance of methane in crystal form frozen at the bottom of the Arctic Ocean and also in coastal waters in the continental US. These crystals

---

[25] "Hydraulic fracturing releases natural gas from shale rocks. A borehole is drilled down to and through the shale, a pipe is inserted and cement slurry is injected around the pipe to encase it. The drill rig is then removed, and the rock is fractured in several stages. Fracturing is accomplished when a gun blasts perforations through the cement encased pipe and chemicals are forced through the pipe at high pressure breaking the rock. The fluid and later the gas flow up the well".

are methane hydrates. In the Arctic the temperature and sea pressure keep the methane in crystal form. In the pressure at the approximate 300 meter depths where they are also found in continental waters pressure preserves them as crystals. Bring methane hydrates to the surface and they turn to a gaseous state and disappear. The methane hydrates are a pure form of natural gas and are believed to exist in sufficient quantity to warrant developing a way to mine them. The US Government has been interested in this for some time. An Act of Congress launched an R&D program a few years ago because of the potential of methane hydrates as a source of energy. The DOE and other agencies, supported by private contractors are doing research required by the Act to learn more about the extent of methane hydrates in coastal waters and to find a way to make them available to us. A report entitled Realizing the Energy Potential of Methane Hydrates for the United States was published by the Committee on Assessment of the Department Energy's R & D Program: Evaluating Methane Hydrate as A Future Energy Source, in early 2010.

A quote from the Summary of the 2010 above mentioned report summarizes briefly the status of the program:

> "Research on methane hydrates to date has not revealed technical challenges that the committee believes are insurmountable in the goal to achieve commercial production of methane hydrate in an economically and environmentally feasible manner. However, many scientific and engineering questions in methane hydrates research remain to be answered before it is feasible to achieve commercial production".

The on-going initial assessment of the scale of the potential commercial development of methane from methane hydrates resources on Alaska's North Slope is not scheduled to be completed for a few more years. Similarly, the demonstration of technical and economic recoverability of marine methane hydrate-bearing sand reservoirs is targeted for 2025. At this time the management of the government's program is not in a position to predict that completing the demonstration will occur any earlier than 2025 so we cannot plan on methane hydrates playing any role in our goal to achieve energy independence over the next several years. But the research to date suggests this source of energy will eventually open new options for us.

How Might the Oil and Gas Situation Change If the Moratoria on Drilling are Removed?

Independent Projections of Oil and Gas Reserves in the United States

Several companies and the Department of the Interior's Minerals Management Service as well as the Department of Energy's Energy Information Agency have been involved in assessing the extent of oil and gas resources available within the United States. These studies have sometimes only provided assessments of availability, while others have also looked at the economic impact of capturing these assets to enhance our economy. The most recent and perhaps the most exhaustive is the study sponsored and managed by the National Association of Regulatory Utility Commissioners. It is described below.

National Association of Regulatory Utility Commissioners (NARUC) Moratoria Study

The National Association of Regulatory Utility Commissioners (NARUC) funded a study entitled Analysis of the Social, Economic and Environmental Effects of Maintaining Oil and Gas Exploration and Production Moratoria On and Beneath Federal Lands, Assessment of the Combined Relative Impacts of Maintaining Moratoria and Increased Domestic Onshore and Offshore Oil and Gas Resource Estimates, dated February 15, 2010. The Study was performed by Science Applications International Corporation (SAIC), a large professional service company providing consulting and research services to governments, and the Gas Technology Institute (GTI), the latter firm having expertise in assessments of oil and gas reserves, under the ground and under the coastal waters. A press release issued by NARUC to announce the study is included herein as Appendix C.

The first paragraph of the Executive Summary of that study is reproduced here because it succinctly describes an important feature of the study and a significant result.

> "In recent decades our nation has restricted a significant expanse of federal onshore and offshore Outer Continental Shelf (OCS) lands from natural gas and oil exploration and production. The federal

government estimates these lands may contain 285 Trillion cubic feet (Tcf) of natural gas and 46 Billion barrels of oil (Bbo) of undeveloped energy resources."

The NARUC study is not based on new field work to support its assessments. Instead it takes estimates prepared originally by government agencies and updates them. Specifically it starts with these:

- A comprehensive review of oil and natural gas resources managed by the federal government ("federal lands") within 11 geologic provinces across the United States—*"Inventory of Onshore Federal Oil and Natural Gas Resources and Restrictions to Their Development", 2008 Department of Interior."*
- 2006 Minerals Management Service (MMS) assessment of technically recoverable oil and gas resources in the United States Outer Continental Shelf (OCS)—*"Assessment of Undiscovered Technically Recoverable Oil and Gas Resources of the Nation's Outer Continental Shelf, 2006;"*
- National Petroleum Council—*Supply Task Group of the NPC Committee on Global Oil and Gas—Working Document of the NPC Global Oil & Gas Study Made Available July 18, 2007* and 2007 Study *"Facing the Hard Truths about Energy";*
- American Petroleum Institute—*"Strengthening Our Economy: The Untapped U.S. Oil and Gas Resources", December 5, 2008*; and,
- U.S. Geological Survey (USGS)—*USGS World Petroleum Assessment 2000.*

To understand what the oil and gas drilling industry, and the government agencies that oversee them, are talking about when they speak of reserves of these resources we need a few definitions. The DOI Minerals Management Service recently renamed the "Bureau of Ocean Energy Management, Regulation and Enforcement" provided the following:

Reserves: The quantities of hydrocarbon resources anticipated to be recovered from known accumulations from a given date forward. All reserve estimates involve some degree of uncertainty.

Proved reserves: The quantities of hydrocarbons estimated with reasonable certainty to be commercially recoverable from known accumulations and under current economic conditions, operating methods, and government regulations. Current economic conditions include prices and costs prevailing at the time of the estimate. Estimates of proved reserves do not include reserves appreciation.

Reserves appreciation: The observed incremental increase through time in the estimates of reserves of an oil and/or gas field. It is that part of the known resources over and above proved and unproved reserves that will be added to existing fields through extension, revision, improved recovery, and the addition of new reservoirs. Also referred to as reserves growth or field growth.

Unproved reserves: Quantities of hydrocarbon reserves that are assessed based on geologic and engineering information similar to that used in developing estimates of proved reserves, but technical, contractual, economic, or regulatory uncertainty precludes such reserves being classified as proved.

Economically recoverable resources: The volume of technically recoverable resources that is potentially recoverable at a profit after considering the costs of production and the product prices.

Undiscovered resources: Resources postulated, on the basis of geologic knowledge and theory, to exist outside of known fields or accumulations. Included also are resources from undiscovered pools within known fields to the extent that they occur within separate plays.

Undiscovered technically recoverable resources (UTRR): Resources in undiscovered accumulations analogous to those in existing fields producible with current recovery technology and efficiency, but without any consideration of economic viability. These accumulations are of sufficient size and quality to be amenable to conventional primary and secondary recovery techniques. Undiscovered conventionally recoverable resources are primarily located outside of known fields.

Undiscovered economically recoverable resources (UERR): The portion of the undiscovered conventionally recoverable resources that is economically recoverable under imposed economic and technologic conditions.

These definitions reveal an interesting feature of oil and gas resource estimates underground or under the outer continental shelves. Note they refer to technically and economically recoverable resource estimates. If the estimates are old perhaps because estimates were not updated because of moratoria, the old resource estimates can be very wrong. Resources shut down by moratoria 20 or 30 years ago have not been reassessed with newer technologies like 3-D seismic imaging. Nor has the technical and economic viability related to recovery of oil and gas with newer technologies been considered, so the old estimates might understate the extent of reserves and what can be now be economically extracted. These underestimates can be large.

You can ask yourself, how can the experts make estimates for oil and gas resources described as "unproved" or "undiscovered".' They do it this way: They start with some indication of the presence of oil or gas. The nature of the geology may initially lead experts to do seismic studies to obtain a picture of the underlying rock strata because the geology is similar to other locations where oil or gas has been found. They may then make an estimate with this rather skimpy basis understanding fully that re-estimates will be needed as data is accumulated. If the find is further developed with exploratory wells and more indications of energy are discovered, the estimates will be further refined and presumably will be of improving quality and accuracy. As more wells are added in a field, and new seismic data added, the geology becomes better understood and the estimates of remaining resources become more accurate. With the passage of time the engineers and technicians also improve their techniques for working each particular field. Time doesn't stand still; we evolve to a situation where the resource estimates reflect improving technology for finding and extracting oil and gas and improving economic feasibility. The estimated resources over time can grow substantially from the earlier estimates. History has shown that in a maturing oil field the amount of oil taken out of the ground plus the estimates of remaining oil (or gas) can be several times the earlier estimates.

Table 6-2 is a summary of NARUC study data which compares the current government (Bureau of Ocean Energy Management, Regulation and Enforcement) estimates of oil and gas resources showing the resources not in areas covered by moratoria with the data and estimates assembled in the NARUC study. The data labeled Adjusted Resource Base includes data assembled by various government agencies that reflects all domestic land and off shore and ignores the moratoria restriction. The study team made the additions shown and finally a total column so you can compare the resources in moratoria areas with estimates for the entire domestic oil and gas resources. As you can see there are significant differences: These data indicate we might have over eight times as much natural gas and roughly twelve times as much oil as the DOE Energy Information Agency (EIA) indicates in areas without the limited access imposed by the moratoria. There are limitations to the quality of the estimates made that lead to this conclusion. These will be disclosed as we proceed. In the final analysis however, you will see the estimates could very well understate the real extent of our oil and gas resources.

Table 6-2
Summary of US Unrestricted Domestic Resources Onshore and on the Outer Continental Shelves (OCS) Oil and Natural Gas Plus Resources Restricted by Moratoria

	EIA Estimates (Includes unrestricted resources not impacted by moratoria)	Government Adjustments to Resource Base (Includes all resources, both included in the moratoria and all other domestic resources)	Additions Made by the Study Team	NARUC Moratoria Study Conclusion. (Assuming All Moratoria Are Eliminated) This is the sum of the government's base plus study team additions
Natural Gas (Tcf)[26]				
Onshore				
OCS				
Total Nat'l Gas (Tcf)	**245**	**1748**	**286**	**2034**
Oil (Bbo)				
Offshore Oil (not incl. AK)			37	
Onshore (only ANWAR)			6	
Total Oil (Bbo)	**19**	**186**	**43**	**229**

This study updates the nation's onshore and offshore natural gas and oil resource base in moratoria and non-moratoria areas and exposes how much of these resources we have available if we have a willingness to go

[26] Tcf means trillion cubic feet and Bbo means billion barrels of oil.

after them so we can start aggressively to pursue this option. The NARUC organization, the consulting and research team it hired to do this important work should be congratulated for their efforts. The study not only exposed what our energy resources might be, but also the economic and social penalties we suffer by maintaining the moratoria. Finally they analyzed the data and created estimates of the economic, environmental and social impacts of dropping the moratoria altogether. We'll get into these impacts in the next chapter of this book.

The methodology that the study implemented included review of the individual oil and gas field histories identified to specific individual geographic areas including:

- Oil and gas technically recoverable on federal onshore lands encompassing 21 basin or otherwise defined areas,
- Undiscovered technically recoverable oil and gas resources on federal lands,
- Technically recoverable resources on federal offshore lands including Alaskan, Atlantic, Pacific and Gulf of Mexico, and
- Technically recoverable undiscovered recoverable offshore oil and gas resources.

Explanation of the Additions Made by the NARUC Study Team

When the drilling and exploration companies decide a new field might contain oil, they do so based on their analysis of seismic data. If they are successful in their initial drilling and conclude more drilling may provide oil or gas or both at costs which meets their objectives, they will proceed. After drilling more exploratory wells they will update their assessments and drilling and exploration plans. If the resources are too meager or too expensive to warrant continuing, they may cease operations and write-off the investment made to date including whatever they paid the government or holders of the mineral rights for the rights to drill in that field. That's the downside. The upside is when they conclude the amount of recoverable resources meets or exceed their anticipated results and they charge ahead.

What is very often found over a period of time and more drilling is the recoverable resource estimates grow. There are several reasons for this. Improvements in technology manifested in, for example, 3-D seismic imaging, help locate oil or gas, and their ability to steer the drill bits resulting in horizontal drilling. In cases like drilling in the Marcellus shale of Pennsylvania, the gas bearing shale might exist in layers only 50 or a 100 foot thick. With vertical wells, the gas production was limited by the vertical distance of the well penetrating only the short passage through the seam of shale. Now, with a horizontal well, gas production is much greater since the well can follow the horizontal shale seam for long distances. The technology to fracture the shale causes the gas to be released in the fractured increments along the horizontal length of the well. These technologies are also used in oil bearing shale as for example in the Bakken Range of North Dakota. An additional technology improvement permits multiple horizontal wells to be drilled from a single platform thereby minimizing the impact of damage to the land in the event of problems at a well site. The drilling industry argued, albeit unsuccessfully when they sought permission to drill in the ANWR, that they only needed to operate from a single platform to drill several horizontal wells to reach much of the area so there was little potential damage to the permafrost.

It is important to realize that a significant tool used by the study team to analyze the data is the DOE's National Energy Modeling System (NEMS). This tool is also used by the Congress and the Administration in their studies of our country's energy issues. Consequently it is safe to conclude that the study team knew what data provides credible results to minimize questions by people who might be prone to doubt their conclusions.

The NARUC study also included nationwide natural gas estimates that were provided by the Potential Gas Committee. The committee is an independent body associated with the Colorado School of Mines which prepares and publishes biennial assessments of the nations' gas resources. These estimates have done so since 1964. Their end of 2008 estimate indicates a resource base of 1836 trillion cubic feet. This compares with the study team estimate of natural gas availability at 2034 Tcf.

How long will 229 billion barrels of oil and 2034 trillion cubic feet of natural gas last? See Table 6-3.

Table 6-3 How Long Will the Indicated Resources Last?			
	NARUC Moratoria Results	2009 US Consumption	Years of Supply at the 2009 Rate of Consumption
Domestic Natural Gas Availability	2034 trillion cubic feet	22.8 trillion cubic feet	89.2
Domestic Oil Availability	229 billion barrels of oil	5.2 billion barrels of oil	44.04

Don't be alarmed, we will not run out of natural gas or oil in the relatively short times indicated in the table. There are several reasons why:

1. It will take time to ramp up leasing, licensing and permitting, exploring and producing on our land and coastal areas so we will only gradually wean ourselves from oil imports.
2. Oil consumptions will diminish as biofuel becomes technically and economically replaceable for gasoline and diesel fuel. We don't know when this might happen.
3. Hybrid automobiles will burn less and less gasoline as battery technology improves and plug in electrical recharging becomes available. Given the relatively slow rate at which the inventory of cars and trucks is updated it might be many years before the battery becomes the primary motive force for our vehicles.
4. Natural gas will replaces some gasoline for cars and trucks thereby extending the years of supply of oil.
5. If methane hydrates become a viable fuel source it may last for centuries.

All of these things will gradually reduce oil consumption used to make gasoline. This will eventually diminish imports but may take years to accomplish. In the end game, we will be best served if most of our cars and trucks run on electricity. In the meantime, we can only reduce imports by increasing domestic production of oil and converting some cars and trucks to natural gas.

Of the five factors cited above, the activity likely to have the most significant impact is number 3, the conversion to hybrid vehicles, but success with hybrids and electric cars is dependent on improving battery technology and particularly the development of inexpensive batteries with high energy densities that can be recharged at home. When that happens we are on our way to the complete electrification and greening of our country.

There is Need for a Comprehensive Survey and Analysis of What US Domestic Oil and Gas Really Exist

Given the way in which oil and gas reserves are estimated as discussed above, we note that estimates have to be built-up for each major geographic area where there is or used to be drilling activity. The locations cover large areas where we now have wells on land and the outer continental shelves or where we once had some prospects for drilling. In many of these areas, moratoria were imposed and further development never happened. We noted, when attempts have been made to make assessments for areas included in the moratoria, the quality of assessments is not good. Other areas where drilling has been permitted, as in parts of the Gulf of Mexico for example, new seismic data and better understanding of the local geology, geochemistry and geophysics have developed and we have much better data with which to make assessment of remaining reserves. Since so much of our potential oil and gas bearing resources have been covered by moratoria, the summary estimates shown in Table 6-2 have to be regarded as weak. History has shown, however, that many early oil and gas plays ultimately have proven to have significant resources so we can probably assume, that many of the areas that have been off limits because of the moratoria, could ultimately provide abundant energy resources.

The NARUC study report contains a recommendation that the federal government should underwrite new 3D seismic studies, particularly of likely oil or gas resources within the moratoria areas, to develop up-to-date assessments of what this county's oil and gas resources really are. In all likelihood the results will provide data indicating that we have availability beyond that shown in Table 6-2. If this is the case, the attractiveness of areas shown to have high probability of extensive resources will result in greater lease prices at the Government's auctions to initiate possible development of the indicated energy resources. In this manner the government may recover some or all of the funds spent for the seismic surveys. It will also accelerate activities leading to increased productivity in drilling and reducing our need for foreign oil purchases.

Recent Event-Drilling In Coastal Waters

As this was being written we saw a massive oil spill in the Gulf of Mexico off the coast of Louisiana. A drilling rig caught fire and sank over a well in the ocean bottom 5000 feet below. This needs to be mentioned because many people are over reacting. This tragic accident is just that, an accident from which we will learn how to prevent future accidents in the same way we have learned from the many aircraft accidents which have occurred in our history. The calls to stop drilling were excessive and counterproductive. It makes more sense to temporarily stop deep water drilling and let the more shallow water drilling, that we have reliably done for years, continue because we need the oil it produces.

Sen. Mary Landrieu from Louisiana has been a beacon of calm and common sense during this crisis. She's stayed focused on the immediate problem, avoided early accusations, and tried to keep the accident in perspective. At a recent Senate Environmental Committee hearing she summarized it neatly.

"I know that this committee has its eyes on the environment. We in Louisiana . . . not only have our eyes on it, we have our heart invested in it and we are making a living on that delta. But we need the oil that comes from offshore to keep this economy moving. We must examine what went wrong, weigh the risk and rewards, fix what is broken and move on . . . If we could do without this oil, we would. But we simply cannot—not today, not in the near future."

Our Federal government, however, has not reacted well. Instead they imposed yet other drilling moratorium in spite of our drilling history in the Gulf of Mexico: "The halt has been mandated despite contrary judicial opinions and a record of nearly 4,000 Gulf wells having been drilled beyond 1,000 feet and 700 wells of 5,000 feet or greater—all without a previous major spill. Further, about a third of U.S. oil production comes from the Gulf, and 80% of that is from the deepwater."[27]

---

[27] The Motley Fool commenting about the ban on drilling in the Gulf of Mexico in a presentation dealing with energy investments. September 2010.

The author of this book lives in South Carolina where one of our politicians was recently heard to utter in responds to the Gulf disaster "No drilling in coastal South Carolina". He needs to be enlightened. The indications of hydrocarbons beneath our coastal waters are suggesting significant deposits of natural gas (and methane hydrates). If there is a spill in a gas well at sea, there is not likely to be near the environmental damage that would be noticed on the land. Further, the State could benefit greatly from a significant find of natural gas.

CHAPTER 7

Economic Impact

THERE ARE SEVERAL reasons to pursue the rational energy policies outlined in this volume, but perhaps the most compelling is the impact on the US economy. Our country is in the midst of a protracted recession that seems to defy the best efforts of government policy makers to turn it around. Unemployment persists at over 9.0 %. It is recognized by many that a few million people have given up the job search. Estimates of real unemployment range from 15 to 20 million people out of work. Many are on the public dole in one way or another, unemployment insurance payments, food stamps or whatever. Regardless of what the numbers are it amounts to a huge number of people who are not paying taxes or making a contribution either to support their families or the general welfare. The government is running seriously in the red. It need not be this way. Implementation of The Rational Energy Policy would put many people to work. Government revenue would increase significantly from tax revenues, individual and corporate, and other payments related to drilling activity. Additional benefits are also described in this chapter.

A few recent studies are described which provide relevant economic impact data.

Price Waterhouse Coopers Study: Economic Impact of the Oil and Gas Industries

We need to begin by examining the contribution to the economy made by the current level of oil and gas industry activity in the US. To that end we look to a recent study by the firm of Price Waterhouse Coopers done for the American Petroleum Institute. Their study is entitled The Economic Impact of the Oil and Natural Gas Industry on the Economy: Employment, Labor

Income and Value Added,[28] It is noted that the study uses 2007 data which was the latest data available to the researchers at the time.

In preparing economic impact studies economists typically look at direct, indirect and induced impacts. The authors of this study define these terms thusly:

1. "The *direct impact* is measured as the jobs, labor income, and value added within the oil and natural gas industry.
2. The *indirect impact* is measured as the jobs, labor income, and value added occurring within other industries that provide goods and services to the oil and natural gas industry.
3. The *induced impact* is measured as the jobs, labor income, and value added resulting from household spending of income earned either directly or indirectly from the oil and natural gas industry's spending".

Their study concludes that the existing oil and natural gas industry employs 9.2 million full and part time people in the US. This amounts to 5.2 percent of the total employment in the country.

This is significant but what could it be if we could convert most of our foreign oil purchases to domestically produced oil?

Since demand for energy grows steadily as do our imports of foreign oil, it is reasonable to assume that a transition to increased domestic drilling on land and off shore, currently forbidden by the moratoria, will eventually involve many more people and expanded capital investment so the economic impact would grow significantly.

The Price Waterhouse Coopers information is presented because it gives us a baseline of economic data.

The president of the American Petroleum Institute recently indicated in an evening news program, that he thought relief from the moratoria could add a million jobs to the economy.

---

[28] Price Waterhouse Coopers. The Economic Impact of the Oil and natural Gas Industry on t he Economy: Employment, Labor Income and Value Added, September 8 2009.

If the moratoria on drilling were significantly relaxed we would expect the economic impact would also increase significantly. In the next few paragraphs we review another study done by Natural Resource Economics, Inc. where they examined the Marcellus shale development in Pennsylvania and a couple of contiguous states[29].

Natural Resource Economics, Inc. (NRE) Study: The Economic Impact of the Marcellus Shale: Implications for New York, Pennsylvania and West Virginia

It has been known for a long time that the Marcellus shale underlying Pennsylvania and parts of New York and West Virginia has natural gas. Years ago there was drilling in these states but it tapered off and the drillers lost interest. Along comes new and more efficient technology in the form of horizontal drilling and shale rock fracturing and the industry's view of the potential of this area has changed completely. The drillers are back in Pennsylvania and West Virginia. Not so, however, in New York State that has its' own drilling moratoria. [All other moratoria referred to in this book refer to those imposed by the Federal Government.]

The purpose of NRE's work was to determine how the economies of these states are impacted by the drilling activity. In the case of New York, however, they only estimate the potential of how it might be impacted if its' moratoria is lifted.

It should be noted, NRE uses the same economic type of modeling tool that was employed by PriceWaterhouseCoopers in the earlier study mentioned and briefly described above. In NRE's case they use a version of it developed by Minnesota IMPLAN Group, Inc.

Among the items found in the documentation is the statement that the three states have natural gas reserves valued at two trillion dollars at current prices (approximately $4.00 per thousand cubic foot). This simply fortifies the evidence presented in Chapter 6 on the availability of oil and gas in this country in voluminous quantities.

---

[29] Natural Resource Economics, Inc. The Economic Impacts of the Marcellus Shale: Implications for New York, Pennsylvania and West Virginia. July 14, 2010

The NRE product includes estimates of impact based on 2009 data. The result of their work for West Virginia and Pennsylvania is a finding of total regional impact of value added of about $4.8 billion. They note that this is spread over a wide swath of the economy generating 57,357 jobs and $1.7 billion in local, state and federal tax collections.

In the October 29th 2010 *Wall Street Journal*, an editorial entitled "Rendell's Frack Attack"_discusses the Governor of Pennsylvania's attempts to extract taxes from the drilling companies. Nothing new there but the interesting part of the article is a short paragraph citing researchers at Penn State University who have studied the economic impact of the drilling in that State. Their findings were that the gas boom will boost the economy by $8 billion and provide 88,500 jobs in 2010. They also indicate this will grow to $14.4 billion and 160,000 jobs by 2015. [Note the Penn State economist's estimates of employment figures for that State exceed the NRE estimates for employment gains in both Pennsylvania and West Virginia. The difference may result because the Penn State people may have more up-to-date data for their State and have a better understanding of the happenings in their home territory. Another reason could lie in the structure and detail of the input-output models they use to do their analysis. Their modeling technology could be better suited to their State. Input output models are a major analytical mechanism for assessing the relative importance of the many variables used to make economic assessments. The relationships between variables are described by mathematical models that may change from time to time as conditions change. Consequently every organization using the basic input output technology for economic assessments may have differences in details depending on their individual applications and the skill and experience of their analysts.

What Cost is New York Paying by Continuing its' Moratoria on Drilling?

NRE made an assessment of what New York State is missing by maintaining their self imposed moratoria**.** To make the assessment a number of assumptions were required. As such they assumed a number of wells drilled and money spent in doing so by 2015 and 2020. Their assumptions were constrained to what could happen in only the counties bordering the northern Pennsylvania border.

Specifically, they assumed drilling 314 and 340 wells in 2015 and 2020, respectively. They concluded that gross regional economic product would be increased by $1.7 billion in 2015 and $1.9 billion in 2020. Employment gains would be 15,727 growing to 18,027 by 2015. Local and State tax payment would increase from $214 million in the earlier period growing to $246 million in 2020. Federal taxes payment would similarly increase. NRE also made an assessment that assumes even more drilling activity with significantly more gains but they are not addressed herein given New York's reticence to allow any drilling. Our report assumes that if New York were enticed to permit drilling it would likely be constrained to a limited area and a limited number of wells. It is believed that New York's reticence about drilling results, at least in part, from a drilling incident where a well was placed in New York City's watershed and some evidence was found of leakage of the chemicals used to support the "fracking" used to release the natural gas from the shale. New Yorkers should remain vigilance of the fracking technology as there is much research underway to find new nontoxic chemicals to replace those found objectionable.

National Association of Regulatory Utility Commissioners (NARUC):

Analysis of the Social, Economic And Environmental Effects Of Maintaining Oil And Gas Exploration And Production Moratoria On And Beneath Federal Lands, Assessment of the Combined Relative Impacts of Maintaining Moratoria and Increased Domestic Onshore and Offshore Oil and Gas Resource Estimates

We referred to this study in the previous chapter because the study provided an update of the domestic availability of oil and natural gas that included assessments of these resources that are not available because of moratoria imposed by the federal government. These new assessments are valuable because oil and gas assessments are normally predicated on the following underlying assumption: They include only what the assessor's believe can be extracted from the ground with current technology at the current prices of oil and gas. Consequently, assessments made years ago before a moratoria was imposed, require updating because both technology and prices have changed considerably over the years. Technology improvements have been made in locating energy resources as well a drilling and extraction. What might have not been economically feasible years ago may be very feasible today. The meaning of all of this is simple. There may be far more oil and or natural gas locked up by the moratoria than we ever imagined.

We turn once again to the NARUC study because it was far more comprehensive than just a study of energy resource availability. It also developed estimates of the economic impact on our domestic economy of not drilling and exploiting the energy resources blocked by the moratoria. It also looks at the social and environmental impacts of not drilling in these same areas. The major tool for developing impact assessments is the earlier mentioned National Energy Modeling System (NEMS) but modified by NARUC. The use of NEMS certainly lends credibility because they are using the government's own tool for impact assessments. In using this tool, the analysts were able to develop three categories of impact assessment; social, economic and environmental. While the study team examined several alternative scenarios, we are interested in the basic conclusion for the alternative the study defines as the "combined comparative case". This provides us with the drift of the results for the entire impact study. Namely there are distinctly negative consequences of ignoring much of our energy resources and instead buying oil and gas in foreign markets.

The NARUC findings for the case selected reveals significant deterioration over the years from 2009 out to 2030; most likely from depletion of the available energy resources. For example they predict a 15% reduction in domestic oil production. They made a similar reduction in natural gas production but the timing of their estimate may have been such that they didn't appreciate the extent of recent experience in drilling in the Marcellus shale mentioned above. The fall off in oil production will also result in reduced economic activity with declines in gross domestic product of approximately $2.36 trillion. As US production decreases, the purchases of foreign oil would increase by about 4 billion barrels over the roughly 11 million barrels of daily foreign oil purchases we experienced in 2008[30]. The billion dollar a day currently spent for foreign oil would continue its' relentless climb.

The US dollar has been depreciating on the world markets making the price of oil more expensive. Some economists are predicting oil will soon

30 It is unclear the period of time over which NARUC expects the 4 billion barrel purchases to be made.

climb to $150 dollars per barrel. Our foreign oil purchases are already climbing as a consequence of the government's response to the BP oil spill in the summer of 2010. Recall their first response was another moratorium on deep water drilling that was ultimately lifted but replaced by more regulation on drillers. In the mean time the operators of deep water drilling rigs packed up their now idled drilling rigs and moved them to places elsewhere in the world where the owners could make money. The US has lost valuable drilling rigs for a while, maybe a long while. The biggest impact in the short run will be increased foreign oil purchases. If the economists predicting higher prices are right we may see the daily cost of foreign oil purchases pushing well over a billion dollars a day. Can we afford it? No! We would be far better off as a country, if we paid our own people to work in an expanded drilling industry and gave the USA a big economic boost.

We noted earlier that we import the majority of the oil we use. Given the growth of the country, and improvements in the economy, we are beginning to see the NARUC's study team predictions of declining production from energy resource areas, therefore we should expect much higher prices for oil and far higher daily cost for oil imports. Wouldn't it make more sense to strengthen our own economy by increasing drilling into our own oil resources?

Northern Economics with the Institute of Social and Economic Research, University of Alaska. *Potential National Level Benefits of Alaska OCS Development,* A study prepared for Shell Exploration and Production, February 2011.

This is a follow-on study to one made and released in 2009, also sponsored by Shell that looked at the economic benefits that might accrue from developing natural gas and oil resources in the outer continental shelves (OCS) in the Beaufort Sea and Chukchi Sea north of Alaska. This study focuses on a more careful examination of the benefits that could be paid to federal, state and local governments over a fifty year span of development and production. The results of this work are summarized herein. [Check a world atlas to see where the Beaufort and Chukchi Seas are located and you might surmise that obtaining permission to exploit these energy resources will be less a problem than coaxing people to work there.]

Over a fifty year period Northern Economics estimated 10 billion barrels of oil and 15 trillion cubic feet of natural gas could be extracted. Employment generated by this work is estimated at a total average annual, direct and indirect, 54,700 people over the 50 years. These jobs would be located in Alaska and the rest of the country. The payroll to compensate these employees they estimated at $145 billion, $63 billion for Alaskans and $82 billion for the rest of the US. The revenue that governments might collect based on oil at $65 per barrel were estimated at $193 billion with $167 billion of this going to the federal government; the remainder is distributed to state and local governments. Since the revenues flowing to governments are mainly based on the prices of a barrel of oil it is interesting to note how the total government revenues increase as oil prices change; at $80 per barrel: $214 billion, at $100 per barrel: $263 billion, and at $120 per barrel: $312 billion.

Government Revenues from Domestic Drilling for Oil and Gas

The US Department of Interior has an Office of Natural Resources Revenue. The Office is responsible for collecting revenue associated with removal of specific minerals removed from federal onshore and offshore lands. They claim these are the largest source of revenue the US government has after the corporate and personal income tax. Data on collections for three years are shown in Table 7-1 below.

Relevant definitions applicable to the table:

- Royalties: The dollars shown for natural gas, natural gas liquids and oil are royalties based on the values of the mineral extracted. The American Petroleum Institute reports that the royalty rates are: onshore wells: not less than 12.5% of the value of production; off-shore wells 18.75% of the value of production.
- Rents: Annual payments by lease holders during the primary lease terms. Rental payments are in addition to royalty obligations spelled out in the lease documents.
- Bonus: Cash paid by successful bidders for a mineral lease.
- Other Revenue: Specified lease obligation revenues, gas storage agreements, invoices, etc.

Table 7-1 Reported Royalty and Other Relevant Revenues			
	Government FY 2008	Government FY 2009	Government FY 2010
Natural Gas	$5,817,764,667	$2,737,482,363	$2,187,464,890
Natural Gas Liquids	$604,749,570	$258,107,132	$419,564,978
Oil	$6,171,889,042	$3,837,936,604	$3,988,484,900
Rents	$304,030,259	$294,325,484	$300,697,800
Bonus	$10,145,062,109	$1,980,925,442	$1,301,971,794
Other Revenues	$11,743,875	$16,266,506	$11,797,250
Total	$23,055,239,522	$9,125,043,531	$8,209,981,612

The three year total for 2008, 2009 and 2010 is over $40 billion. Can you imagine how these revenues might grow if our oil companies were allowed to drill in areas now restricted by the moratoria?

In addition to the fees described above that the oil companies paid to the government, they also paid federal income taxes, state income taxes, property taxes and foreign income if they operate overseas, as most of them do. The DOE's Energy Information Agency maintains a Financial Reporting Service (FRS) that compiles and consolidates data on the major energy companies operating in the US. There are approximately 30 companies whose data is included. These companies accounted for 43% of the total production of oil and natural gas liquids and 43% of natural gas production. The EIA data indicates that for the nine year period including 2000 through 2008 the FRS companies' actual payments to the IRS were $285 billion.[31] The American Petroleum Institute reports that the total income taxes paid from 1980 through 2008 reached over a trillion dollars, and, in 2008 alone the amount paid was over $20 billion.

These data clearly show that the federal government would be a major beneficiary of increased drilling for oil and gas. So too would many of the states and local communities that would supply the labor and materials and services as the drilling and related industries expand as moratoria are

---

[31] DOE EIA Financial Reporting System For 28 Schedule 5112 Analysis of Income Taxes

eliminated or modified to permit more drilling. Perhaps some of the tax revenues generated could be used to pay down our atrocious federal debt.

Conclusion

The Price Waterhouse Coopers Study showed we have 2.1 million people directly employed in our oil and gas industry rising to about 9 million when the numbers of people indirectly involved are added. These numbers relate to a situation where we are producing less than half of our needs. Imagine how the numbers of people employed might grow if we went all out to cut imports by increasing our domestic drilling wherever we can find it. It could employ many more people and provide economic growth and significant growth in government revenues. The economy of the entire US would benefit from building our oil and gas industries, but so too will government revenues. The amounts flowing to government could go a long way towards solving our state's and the national debt problems. Growing our oil industry may be the only way we have to deal with our country's debts.

Economic Impact of New and Modernized Electric Power Generation Facilities

It is impossible to predict the total economic impact of what has to happen to implement the conversion of the hundreds of coal burning power to natural gas and ultimately to nuclear power or fuels that emit no noxious products or green house gases. The reason is the unpredictability of what the utilities will do to modernize their facilities and then begin the long process of planning, evaluating choices, permitting and finally construction. Compounding their problem is predicting future electricity requirements in their respective areas of interest. For example, the increasing use of all electric vehicles will draw power from their home electricity supply increasing demands on the power companies. Another consideration some utilities must deal with is the possible need to convert sea water as fresh water supplies become insufficient given population growth. There are just too many variables to permit making any sort of meaningful economic impact assessment.

There is some data available, however, where major utility companies are already involved in the complicated planning process and have made economic impact assessments. These provide insight into the estimated size of the impact in their areas of interest.

Dominion Resources is the major utility company serving the needs of the State of Virginia. In a recent press release, it provided these data relative to ongoing and planned improvements in their systems.[32] Nine ongoing and planned power generation and environmental projects will provide more than $3.3 billion in benefits during the time span between 2007 and 2015. The projects will support 14,200 construction jobs in this period and thereafter $290 million in annual benefits to Virginia and 750 jobs beginning in 2015. None of the nine projects involved nuclear power stations. The economic impact assessment was provided by an independent consulting firm.

Had Dominion Resources work included new nuclear plants the dollar values and schedules would be much larger. For example, another big power company, FPL in Florida is seeking to add two additional reactors to their Turkey Point site where two reactors are already in operation. They indicate the project will provide $6 billion in economic benefits spread over ten years and create 3,600 jobs during construction followed by 800 permanent jobs when the plants become operational.

These two power companies and their situations are normal and will be expected to one degree or another all over the country. Recall the data provided in Table 3-1 on the numbers power plants in the US fueled by coal, oil and natural gas. These were respectively, 1,445, 3768 and 5,569. The coal and oil fueled plants should be converted to natural gas in the short term and later, all of them should be converted to non emitting nuclear, wind or solar. Extrapolating the economic impact assessments we have for Dominion Resources and FPL suggests enormous impacts over many years across the country.

---

[32] Press Release entitled Dominion Virginia Power's Construction Program Brings Virginia more than $3 Billion in Economic Benefits dated August 15, 2011 /PRNewswire.

CHAPTER 8

Recommended Energy Policies

Introduction

THIS CHAPTER SUMMARIZES important findings presented in this book and defines the energy policies we advocate. Reasons to support our positions are presented. It is followed by a discussion of what the government has to accomplish to facilitate implementation of the rational energy policies presented. Finally there is a short discussion of what citizens should do to support the adoption of the ideas expressed in this book.

Objectives of the Rational Energy Policy

We begin by repeating the objectives recited in Chapter 1 and then we expand on them with insights developed along the way.

1. Energy independence: Minimization and ultimate elimination of the over $1 billion per day we spend for foreign oil.
2. Cleaner environment: Minimization and ultimate elimination of manmade noxious and unnecessary greenhouse gases from transportation and power generation activities.
3. Economic expansion: Significant creation of permanent jobs.
4. Strengthen our economy: Grow government revenues so officials, hopefully, will pay down the national debt. Reduce the trade deficit: Good for the economy.
5. Increase national security by minimizing dependency on foreign oil and possibly minimizing the size of our standing military

forces. Increase national security by minimizing dependency on foreign oil and possibly minimizing the size of our standing military forces.

Energy Independence: The majority of our oil is purchased from other countries (Approximately 11,000,000 barrels per day). The recession experienced in the last few years by the US with its' decline in manufacturing activity has reduced our oil requirements and the amount of oil purchased. The recession has been a global one and so the decline in oil usage has resulted in reduced global demand and has provided a period where a barrel of oil has cost less than it would have had global demand been more robust. Now in early 2011, the US and much of the rest of the world has begun an economic recovery and oil demand is rising and so are prices. In addition, a new phenomenon is occurring. Some of the oil producing countries that provide much of the world's oil are having severe political problems that could jeopardize their oil businesses putting further pressure on prices. Today (May 2011) global oil prices are about $100 per barrel and rising; perhaps significantly further.

Energy security is similar to energy independence, but it is really different. Energy security is ordinarily not a problem because there are many places in the world market where we can buy oil so it's always available. But in recent years one of our historical sources, Venezuela, is being cut off by an unfriendly dictator. Now the Mideast sources are in trouble and our usual sources in that corner of the world could also be denied to us. What does all this mean to us? Energy security or having enough oil to meet the demands of our military and our economy may become a real problem. The soaring prices of oil can send the billion dollars a day paid for foreign oil headed skyward. If just paying the daily prices isn't bad enough, it could kill our economic recovery. We are very vulnerable and not as secure as we ought to be.

Clean Environment: There is an ongoing effort by the automobile companies to develop hybrid cars with the best available batteries they can find. They are likely motivated by the recent act of Congress raising the CAFÉ auto mileage standards that become effective in a few years. Furthermore they are certainly aware that gasoline is becoming increasingly expensive and expect their customers to demand better gas mileage. Their first reaction is the move to hybrids, but sooner or later they will want to create all electric cars that can be recharged at home and elsewhere at battery recharging stations. We too should want electric cars because there is no pollution damage to the environment[33]. There are technical issues however that will inhibit an early movement to these all electric cars. First batteries with sufficiently high energy densities haven't yet been invented. Further, rechargeable batteries may take hours to recharge. That's OK if you're recharging in your garage but not so good if you're on the highway. Technically acceptable battery development may be a long time coming because, when developed, they must also be at prices customers are willing to pay. At this time, batteries are expensive. Further battery costs are not likely to experience significant price drop as production ramps-up. The reason is batteries are chemical devices sometimes containing minerals that are fairly rare, hard to mine and expensive. They probably aren't likely to experience the steep learning curves typically experienced with electronic or mechanical devices. The technology issues will one day be resolved, but it will take years to convert the inventory of (over 220,000,000) registered vehicles, since their numbers greatly exceed the numbers of new vehicles sold every year.

The development of new biomass fuels, particularly cellulosic ethanol, are experiencing their own technical and economic issues, so the future for this alternative to oil based fuel is also uncertain.

Recap: A clean environment is a fine objective but we must be realistic about what is achievable. The facts however speak for themselves: Namely we do not yet have an economically viable fully electric car since the technology needs more development time. Once we have one, it will take years to replace all the cars and trucks in the inventory of registered vehicles simply because the inventory is large and the turnover rate is relatively small. Consequently, we will need a lot of oil to power our vehicles for many years. Should we buy it or drill for it domestically?

Economic Expansion: The US is in a period of protracted recession or slow growth. We are also seriously in debt. The interest charges on our

[33] The power generating source for the electric vehicles could be a polluter if it hasn't been converted to nuclear, solar, wind or hydro power.

debt, large amounts of which are being paid to foreign bond holders, China and so on, are huge and growing rapidly. Those bills for interest expenses are money that is not available for our economic development. Without growing economic development we face declining standards of living.

Strengthen our Economy: Our country has been experiencing a serious recession; in fact the worst one since before the Great Depression. Our government has tried to deal with it with little success. In doing so it provided massive spending programs. You can believe they have been successful or not but, the fact remains: we are left with massive debt amounting to trillions of dollars and the economic growth we might be having is not robust. A way to deal with this is to adopt economic growth programs to maximize employment so we have fewer people getting support from the government and more people paying taxes. The oil and gas industries typically pay lot of taxes. Since their profitability is dependent on the price of oil these companies should be expected to pay a lot of taxes given the rise in the per barrel prices of oil. They also pay large sums for leases and drilling rights and they pay again based on the energy resources they extract. Since we have large domestic supplies of oil and natural gas we should go all-out to grow this industry.

Reduce the Trade Deficit: Our trade deficit has been chronically out of balance for years. We can repair it a bit by reducing foreign oil purchases, but we can't fix it completely.

National Security: As pointed out above, our energy independence might be threatened because of possible dwindling foreign oil sources. Some Mid Eastern sources are having political problems of uncertain outcome as well as Venezuela's threats to cut us off. There is another security problem that costs us a great deal. That is the requirement to staff and equip significant military forces to protect our Mid East interests. We have had a fairly large military force stationed there for about a dozen years. We have also lost a lot of people in the fighting in Afghanistan and Iraq. While our politicians predict we will be out of there one of these days, few people will bet on it. In the meantime, it costs us a huge amount of money to support our military forces to protect the flow of oil from the Mideast.

All of the preceding discussion leads us to the following:

Recommended Energy Policies

The recommendations are described in three sets. The first relates to transportation related activities and the second relates to electric power generation. The first, of course, mainly involves oil. Transportation uses consume over 70% of the oil we produce and buy[34]. Electric power generation is provided from multiple energy sources but fortunately relatively little oil. The problems with power generation are costs and unwanted emissions. A third set of recommendations are also included suggesting basic research for which the Government has excellent resources. Some of our recommendations may be more a continuation of the on-going research activities being accomplished or sponsored by DOE and perhaps other agencies as well.

Transportation Related Policy: The United States should adopt the following policies.

1. **Oil and Natural Gas Assessments**: Bring assessments of domestic oil and gas resources up-to-date. We need to really know what oil and gas resources we have and where they are located.
2. **Replace Drilling Moratoria with Less Restrictive Regulation**: Eliminate the drilling moratoria and apply new criteria that will result in minimal restrictions on areas in which drilling will be permitted.
3. **Federal Agencies Should Reengineer and Reassess Regulatory and Permitting Processes**: The government agencies involved in permitting and oversight of drilling activities should revaluate their requirements and procedures to accelerate the growth of domestic exploration and production of oil and natural gas on land and the continental shelves, including all areas of Alaska. In view of the large daily cost of buying oil we should give the processes of permitting and producing more a higher priority.

---

34 The remainder of the oil we use is for industrial purposes not pertinent to our policy recommendations.

4. **Support Use of Natural Gas as Transportation Fuel**: The use of natural gas to fuel commuter's cars and vehicles normally operating in urban environments should be encouraged. The Federal government and states should press for this policy and (a) convert their fleets of vehicles to natural gas and (b) make their fueling stations available to the public.

Electric Power Generation Policy: Our policies should be to:

1. **Support Nuclear Power**: Encourage power generators to maximize their use of modern nuclear power reactors having design features that increase safety. They should look at Small Nuclear Reactors (SNRs) that may be factory built and which facilitate expansion over time in ways to reduce interest expense and capital costs. Nuclear power provides the least expensive kilowatt-hours costs necessary to support growth of manufacturing industries.[35] [The kilowatt-hour costs of a power generator by solar and wind systems are too high to support a manufacturing economy, but nevertheless such power is useful on a limited basis in many locations].
2. **Wind, Solar, Hydro and other non-Polluting Power:** Power generators should be encouraged to further develop wind, solar and hydro power in locations compatible with the technology and where the power can be integrated into the grid and local power company requirements.
3. **Natural Gas**: Natural Gas should be used for conversion of coal burning generators because of relative cleanliness and low cost. This is an ideal fuel to use as an interim fuel until cleaner nuclear power can be made available and cost competitive alternatives become available.
4. **Recycle Spent Reactor Fuel Rods**: The spent fuel rods should be recycled since they have much energy that can be recaptured.

[35] The Three Mile Island accident of many years ago was due to human error. There were no deaths, no injuries and no radiation leakage. Moreover, current reactor designs employ back-up passive fuel cooling systems that would have prevented the problems the Japanese incurred when the Tsunami destroyed the backup power systems on their reactors. As this is being written we have no indications the earthquake preceding the tsunami was the cause of the failures experienced.

The process greatly reduces the volumes of radioactive waste. Some waste products are relatively short lived and only a small amount (approximately 20%) are long lived. These latter products must be stored as was once planned for the Yucca Flats facility. [The original decision by Pres. Ford to not recycle was made because he feared the small amount of plutonium in the spent rods might fall into the hands of terrorists who would make bombs. There are over 250 nuclear power electricity generating plants outside the US, all of whom, except a few in Japan, recycle their fuel rods. Plutonium loss has never been a problem.]

Basic Research

1. Government should fund basic research in battery technology that could contribute to improved designs of hybrid cars and trucks and share the results with battery and automotive interests. The focus should be high energy density, rapid recharging and low cost.
2. Basic research is also needed to facilitate use of electricity made by large scale solar systems and wind farms. The requirement is large battery systems that could store electricity made by these installations until the utilities need it. Mother Nature is not necessarily in sync with the power companies. The wind doesn't necessarily blow when we need it, so power generated is transferred long distances with big losses or the wind farm may be turned off. Wind power is also expensive. Solar power can only be effective in certain areas and it is also expensive. Still another objective of research with wind technology is to find ways to reduce the cost of the power generated. The relatively high kilowatt hour costs of the power generated will inhibit the potential acceptance of wind as a power source. Whether additional basic research in solar technology for power is needed is conjectural. Significant advances in reducing the cost of power with photovoltaic are being made in the US and China.
3. Private interests are doing research and development. Their efforts could be facilitated if test facilities could be made available to them when needed. Such facilities can be expensive but the cost often cannot be rationalized because the need is infrequent. There is significant precedent for government to own test facilities and rent them to corporations from time to time. For example the

government owns a major wind tunnel at Tullahoma TN that it rents to aerospace companies. Similarly, it rents its space launching facilities at Wallops Island, Cape Kennedy and Vandenberg AF Base to aerospace companies.

4. The electricity grid is inefficient, vulnerable and needs several improvements. The problems are widely recognized and are being dealt with on a piecemeal basis. Utilities owning a piece of it are making improvement but a nationwide backbone operating at very high voltages to limit energy losses does not appear likely any time soon. The government should support the industrial development of the backbone when the industry has defined the project and how the government can help.
5. When electricity is transferred over the grid there are line losses amounting to a few percentage points. The lost power costs us the fuel to create that power. If it is made with carbon based fuels, we also generated waste products nobody wants. Use of superconducting cable that can transmit electricity without loses are technically possible but is prohibitively expensive. Basic research in this area should focus on ways to bring costs down. If we could create relatively inexpensive super conducting cable we likely would need to place it underground where it would be easier to keep the cable cold. This will have esthetic appeal to most everybody since nobody likes to look at the miles of high voltage lines that crisscross the country.

CHAPTER 9

Government Actions Needed to Support the Rational Energy Policies

Introduction

THE DOE'S ENERGY Information Agency published its Annual Energy Outlook 2011.[36] Their projections indicate where something might be hindering implementation of the Rational Energy Policy. For example, you will find the following predictions in the 2011 Energy Outlook looking out to 2035.

- Fossil fuels, particularly coal, continue to provide most of growing fuel requirements.
- Nuclear power usage, as percentage of total requirements, hardly increases at all.
- Use of renewable fuels does increase but not a great deal.
- Capital costs increase for all kinds of power generation except wind and solar.
- In terms of capacity additions, the existing 104 nuclear power generating facilities only grow by six units.
- The price of a barrel of oil increases to about $125, however, the range of possibilities range from a low of about $50 to a high of about $200.

[36] Annual Energy Outlook 2011, Reference Case, Energy Information Agency, December 16, 2010

- Crude oil imports decrease only modestly from current levels although consumption of oil also decreases somewhat.
- Ethanol from corn in 2035 is still used and in larger quantities than we use today.
- The number of registered drivers' increases from 207 million today to 265 million in 2035.
- The efficiency of "vehicle stock" they indicate is 20.8 miles per gallon today but growing to 27.8.
- Even by 2035 they still anticipate most cars will be gasoline powered and while the use of hybrids doubles or triples compared to today, there are relatively few plug-in hybrids and all electric cars.

All-in-all, this is not a pretty picture. Clearly we must do more, much more, to change our policies to guide us to a more satisfactory future and which meets the objectives indicated in Chapter 1 and further discussed in the preceding chapters.

A list of the items that the government must do if the Country is ever to receive the benefits we have outlined in this volume is provided with a discussion of each item.

The impediments are listed that must be addressed.

The Oil and Natural Gas Drilling Moratoria Needs to be Re-Examined and Largely Eliminated

1. The moratoria banning drilling on land and sea are sweeping in their geographic coverage and are denying us much needed oil thereby forcing us to buy increasingly expensive foreign oil.
2. The criteria for imposing moratoria should be reassessed and greatly limited in geographic scope. The criteria should include a balance between preservation of nature's flora and fauna with the needs of our immense economy and its needs for mineral resources. The impacted states should be given a priority position in decisions about moratoria.
3. The regulations regarding drilling need to be reviewed. Experts in the industry need to be involved in the review and changes need to be made, not to make them less effective, but to make them user

friendly and to develop a consensus among the drilling community that they are useful and necessary.

4. The inventory of energy resources, particularly for areas that have been restricted by moratoria (both on land and off-shore) should be surveyed with the latest technology to update the data so we can better understand what we have and where it is. These data should be made available to the energy industry and the states. [The updated survey data will likely show more oil and gas because the surveys indicate the resources that can be recovered with existing technology and economic conditions. Since drilling technology has improved greatly over the years and oil is now much higher priced than when the moratoria were originally imposed, the auction prices for drilling rights will likely bring higher prices than otherwise. The government may easily recover its' investment in sponsoring the surveys.]
5. The impacted states should share in royalty payments that now go to the Federal government.

Government Needs to Be Oriented to the Long Term Public Good and Less So to the Passions of the Moment When Acts of God or Accidents Happen

When accidents happen as with the BP deep water oil spill in the Gulf of Mexico in the summer of 2010, the public press proclaims a major disaster with damage to the environment, fish, beaches, and shore communities. The government immediately overreacted and began looking for somebody to hang and closed down all deep water drilling throwing thousands of people out of work. Before they opened the Gulf again to drilling, they promulgated new regulations to make things more difficult than they already were. The new regulations were a product of the government made without input from the people who have to implement them. In the meantime the owners of the very expensive deep water rigs moved them somewhere else on the globe where they could begin to make money.

6. The Nuclear Regulatory Agency should examine its review and regulatory procedures in conjunction with the representatives of the power and nuclear reactor industry and the public it serves to try to find ways to shorten the processes. The lengthy processes

today contribute to the very long cycle times to bring new nuclear facilities on line.

7. Similarly, the National Ocean Engineering and Regulatory Agency should reexamine its recent tightening of ocean drilling in the wake of the 2010 BP oil spill in the Gulf of Mexico. It appears that the regulations serve the purpose of discouraging drilling and not the purpose of only making drilling at sea safer.
8. Governments should move to minimize the time required for licensing and regulation wherever required.

The Federal government should enlist the aid of those states in managing and facilitating the rapid exploitation of our national energy resources. It should also share with those states royalties that would ordinarily accrue to the federal government.

Finally it is emphasized neither the Congress or the White House are likely to act on anything recommended in this volume because of the vested interest of many parties whose actions (read this as donations) have got us in the situation we are today. We need the White House and every congressperson and senator to hear from their constituents, repeatedly, until we see concrete action to implement the policies espoused in this book.

How the States Can Support the Rational Energy Policy

The states can do much to further the objectives of the rational energy policy. For example, the states typically own many cars, light trucks and buses that are available with engines fueled with natural gas, or can be converted to natural gas. Many fleet operators such as utilities have already made the conversion. Specially designed natural gas fueling stations are required. These should be made accessible and conveniently available to the general public. If the fueling stations are available to the public they will be more likely to buy cars fueled with natural gas to commute to work. When that happens there can be significant quantities of people shedding gas guzzlers for natural gas fueled vehicles. Local businesses employing trucks and cars could also make the conversion. With increasing numbers of fueling stations available to the public and the movement to natural gas accelerates, real progress can be made in reducing the need for oil.

It is expected that initially state or local governments may have to invest in creating natural gas fueling stations accessible to the public. These investments can be recovered by selling the stations to private operators as soon as investors can see the viability of the business. There are private companies already which might be attracted to make the investments themselves if the potential is shown to them.

Because of the abundant supplies of natural gas being found in the US, the price of natural gas is no longer tightly tied to oil prices. We can expect oil and gasoline prices to continue a long term trend upward because of the growing economies of developing nations. Natural gas prices should not increase as steeply as gasoline because of the expanded domestic supply so fueling natural gas equipped vehicles may become relatively less expensive. This would induce more and more people to buy these vehicles.

CHAPTER 10

Closing Comments

MY INTEREST IN writing this book has been to convince you that our country would be far better off drilling for oil rather than buying it on the world market. Domestic drilling will create jobs when we badly need them and grow our economy. We need oil today and will for as many years as it takes to convert 220,000,000 cars and trucks to hybrids, natural gas and ultimately electricity. These will become progressively less polluting until one distant day cars and trucks will be absolutely clean with no pollution and no green house gases.

Similarly, our electric power industry today uses too much dirty fuel to energize some of its power plants. We can clean it up but when we do the price per kilowatt hour can go up significantly. Sometimes it goes up too much to provide the cheap electric power that manufacturing industries need. Labor costs may be high in factories but expensive power may provide the double whammy that causes the bosses to move production elsewhere. We can't have any more of that.

Clean power is provided primarily by nuclear energy, solar and wind. Hydropower from falling rivers and geothermal are power sources that also offer clean power but are of limited significance. Wind and solar are great but for the foreseeable future are too expensive and give electric utilities problems. Lack of large capacity batteries limits the capability to use this power when they need it most. Advancing technology may one day make these more attractive in an economic sense, but for now they are too costly to support activities that need cheap power. Nuclear power offers the least costly power even after amortizing the high capital cost to build the power plants. Even here there are events unfolding that offer promise of relief from escalating costs. Small modular reactors are being created in standardized designs, factory built, with minimal on-site assembly. Individual modules are expected to be priced at affordable levels that can be built with others added when needed. Cash flow from the first units built can help finance the purchase and installation of later units. Only nuclear power can provide large amounts of cheap electricity to sustain our growth and manufacturing. Given the time requirements to design, permit and build new nuclear facilities, we should convert dirty coal burning power plants to cleaner natural gas and on some distant day convert them to nuclear power. The ultimate conversion to nuclear will eliminate carbon dioxide emissions that some people contend is a cause of global warming.

The federal government has done the nuclear power industry a vast disservice over the years. It has been charging them a fee intended to fund a repository for spent nuclear fuel rods. The fee has aggregated to billions of dollars. The repository has been built but now the government prevents its' use. A repository, however, is not an efficient way to handle the spent fuel rods. The US is the only country that does not recycle spent fuel rods[37]. The process recovers unused energy in the spent rods and reduces the volume of highly radioactive waste to a much smaller amount. This waste must be placed in a repository, but a much smaller one than what the US had planned. [Yucca Mountain has the capacity to hold all of our high level waste from recycled fuel rods for a very long time.]

Japan, currently dealing with a nuclear problem caused by the combination of an earthquake and tsunami, had spent fuel rods on site. These spent fuel rods have caused a lot of concern for the Japanese authorities causing us to wonder if having the spent fuel rods on the site for our 104 operating nuclear reactors is really as safe as we have assumed. We have the capability to recycle spent fuel rods in a DOE facility in South Carolina. It is already successfully recycling old nuclear warheads into reactor fuel rods.

---

[37] We note Japan recycles spent fuel rods but, in the case of the Fukushima reactors was storing the spent rods at the power plant pending shipment to their recycler. The stored rods contributed to their problems after the earthquake and tsunami in 2010.

The Role of Conservation

We have not explicitly mentioned the role of conservation as a way to reduce energy consumptions but major conservations efforts are implicit in significant ways. The work going on, for example, to improve batteries and create efficient hybrid and electric automotive technology is a major effort to conserve oil. Similarly, the electric power industry's efforts to move to cleaner sources of power preserves the environment. Their efforts to improve the electrical distribution grid could ultimately minimize the power line losses that results in utility companies creating a lot of power later lost as heat to the environment. Are these not conservation efforts?

A Word of Caution

There are a lot of uninformed people in this world. A lot of them work for newspapers and other media. Some of them are also elected officials. The reaction of some of these people is to display their ignorance about accidents leading to disasters by over reacting before the facts are collected and analyzed. We don't close down Boeing when one of its' aircraft crashes but we did close down deep water drilling in the Gulf of Mexico after the BP accident, in spite of the fact that many deep water wells have been drilled without incident. The calls to end drilling for oil or close our nuclear power plants whenever an accident happens only serve to incite people and precipitate poor decisions. The calls to stop building nuclear power plants, ignores the fact that in the 40 or so years, since we added to our inventory of nuclear power plants the nuclear industry has built many reactors for our military and foreign customers. The technology developments accumulated over the years are significant and make them much safer. It also happens that the US Nuclear Regulatory Commission has required our nuclear power plant operators to retrofit their systems with safety improvements.

Now I Have a Question for My Readers

In 1942 President Roosevelt created the Manhattan Project to harness nuclear power. The Project created a nuclear weapon that in 1945 ended World War II. A tremendous accomplishment in three years!

In the late 1950s the Russians launched Sputnik and demonstrated nuclear capabilities. This made the US realize we were behind and the Russians had a strategic advantage in the cold war. To recover the advantage, we commissioned the Navy to create a nuclear powered submarine strategic defense capability that was to be armed with intercontinental ballistic missiles and nuclear warheads. A lot of tough technical achievements were required, but the Navy performed brilliantly and created a force that has served us well for the past 50 plus years. Another tremendous accomplishment in four years!

John F Kennedy, shortly after becoming President in 1963 challenged the country to put a man on the moon before the end of the decade. We did it. Another great engineering achievement!

Given our demonstrated capability to perform highly technical projects in very short times, why then can't our government now organize itself to get things done expeditiously? Why does it take years to get leases and permits to drill for oil on federal properties on land and on the continental shelves? Why does it take years to bring a new nuclear power plant from the point where a need is recognized and the government issues the appropriate permits to build and operate the plant? Surely the forces of industry and government could get together and do some process re-engineering and find better and faster ways to get the job done. They have to stop acting like they are on opposing teams. They should realize we are in national emergencies; one to grow our economy and save the billions spent buying foreign oil and another to modernize the electric power industry to make it cleaner, safer, more efficient and offering electric power at rates supporting manufacturing jobs.

The government has to get behind the effort to bring these policies to fruition although it seems too much of the government is captive to single interest forces that will resist change of any sort. But in the final analysis the Congress and the White House will be responsive to the voters. If enough of us start calling, writing or emailing the White House and our elected officials in Congress to tell them you want the rational energy policies defined in this book, maybe we can make some progress. Contact them often until we get results. Don't accept being ignored.

Thank you for reading this book. I hope you have learned something useful.

APPENDIX A

Summaries of Opportunities to Reduce Use of Gasoline or Diesel

As Fuels for Automobiles and Trucks

Table A-1 Summaries of Opportunities to Reduce Use of Gasoline or Diesel As Fuels for Automobiles and Trucks

Fuel/ Power Plant Technology	Status	Problems	Prospects	Conclusion
Corn Based Ethanol Supplement To Replace Gas	Corn based ethanol production is subsidized (over $0.50 per gal) with quantities increasing each year.	Corn prices have been driven up negatively impacting food prices. At best, ethanol offers a small amount of energy over the amount of energy required to grow corn and make it.	Hopefully corn based ethanol will be replaced by cellulosic based ethanol over the next couple of years. The subsidy should be eliminated soon.	Ignore it since it should soon disappear. If it doesn't it is because Congress in more interested in appeasing constituencies than efficient use of the public's money.
Cellulosic Ethanol Supplement To Replace Gas	A lot of private and government money has been poured into the development of the technology to make ethanol from grasses, forest and farm waste. Ethanol has been produced from prototype development plants, but yet from production facilities. Much better energy situation than with corn based product.	We need a production capability to mass produce the ethanol at low cost. Ethanol is a cumbersome product because it cannot be carried in pipe lines so must be blended near points of distribution.	Cellulosic ethanol can probably become abundant today if cost is low enough so a larger percentage might be used in cars. Several production plants are under construction with planned opening by 2011	We need cellulosic ethanol as a bridge technology pending ultimate replacement by electrification. Cellulosic facilities emit less GHG than corn based manufacturing because they are not powered with CNG.

Table A-1 Summaries of Opportunities to Reduce Use of Gasoline or Diesel As Fuels for Automobiles and Trucks—continued				
Fuel/ Power Plant Technology	**Status**	**Problems**	**Prospects**	**Conclusion**
Biodiesel	A vegetable oil-or animal fat-based diesel fuel typically made by chemically reacting vegetable oil, or animal fat with an alcohol. Most use is in the EU as a blend (20% +). Soybeans are a major source of vegetable oil used for biodiesel.	Using the current yields, vast amounts of land and fresh water would be needed to produce enough oil to completely replace fossil fuel usage. It would require twice the land area of the US to be devoted to soybean production	Biodiesel shares many of the problems of ethanol production, particularly in land use. One projection is that it could displace 20% petroleum, but not much more.	Biodiesel seems less a promise compared to natural gas for trucks.
Other Biomass Fuels	This type of fuel has most of its current interest in algae, particularly marine algae. There is much interest and possible future progress.	This area is at a very early stage of research and development. Only small scale work has been accomplished.	This topic should be encouraged as having a significant future potential.	Watch and wait. Encourage academic and private research and development.
Natural Gas	Natural gas has been in use as fuel for both trucks and cars for some time. No new technology is required. New drilling technology to extract "shale gas" is making natural gas more abundant and cheaper. Note the historical relationship between the price of oil and natural gas is disappearing because natural gas is more abundant and domestically available. On the other hand, world oil prices are driven by the OPEC cartel.	The major impediment to the growth of natural gas has been gas prices and the lack of fueling stations available to the public. Natural gas is not a good fuel for interstate truckers. Congress' problem is how to incentive the situation to make people build Natural gas fueling facilities and people to then buy natural gas powered cars and trucks.	If many of the areas where drilling is forbidden become open areas, even more natural gas should become available (and cheaper) so use as transportation fuel may become very reasonable.	The US should encourage use of natural gas as transportation fuel. Since no new technology is needed and conversion of some existing vehicles is possible we will be able to see results sooner than any other alternative.

Table A-1 Summaries of Opportunities to Reduce Use of Gasoline or Diesel As Fuels for Automobiles and Trucks—continued				
Fuel/ Power Plant Technology	**Status**	**Problems**	**Prospects**	**Conclusion**
Hybrid Gas With On Board Chargeable Battery	Several car manufacturers offer hybrids and there product offerings improve each year.	Battery technology limits the performance of the hybrids so eventually better batteries with high energy densities, long life and low cost will replace the current product offerings		Let the industry mature on its' own.
Hybrid Gas With Plug-In Rechargeable Battery	Several car manufacturers will soon offer hybrids with rechargeable batteries. The cost of batteries may inhibit the sale of these cars. The incremental cost of a hybrid car has to be less than the expected fuel savings to induce the maximum number of purchaser. If it isn't only buyers willing to pay the price for the novelty will buy them	Homes need "smart electric meters" to benefit from lower evening rates and recharging stations in their homes. Publically available charging stations are needed to support the traveling public.	Depends on the price, manufacture warranty, battery replacement cost and frequency	The government may have to offer incentives to cover excess costs until prices drop.
Fuel Cells	Several vehicle manufacturers or fuel cell suppliers are working to develop a workable solution.	Fossil fuels can be reformed to extract the hydrogen fuel but by products such as CO2 are a problem. Electrolysis can be used to provide hydrogen but the utilities may be required to increase capacity.	To soon to predict how this development activity will evolve.	

APPENDIX B

Descriptions of Selected Fuel Sources

Used to Create Electrical Power

Table-B-1 Descriptions of Selected Fuel Sources Used to Create Electrical Power	
Fuel Source	**Process Description**
Coal Supercritical	Supercritical coal plants employ advanced technologies to achieve higher temperatures and pressures to obtain high thermal efficiencies and higher power output than earlier coal fired steam driven generators.
Coal Integrated Gasification Combined Cycle (IGCC)	IGCC technology turns coal into "synthesis gas" where impurities are removed resulting in low levels of sulfur dioxide and mercury. The gas is combusted to drive a generator. Waste heat is used to drive a steam cycle to create additional power.
Coal IGCC with Carbon Capture and Storage (IGCC with CCS)	Same as above but carbon dioxide is pumped into deep wells where it will stay forever.
Gas Combined Cycle	A combined-cycle gas turbine power plant consists of gas turbine generators equipped with heat recovery steam generators to capture heat from the gas turbine exhaust. Steam produced in the heat recovery steam generators powers a steam turbine generator to produce additional electric power.

APPENDIX C

Press Release

NATIONAL ASSOCIATION OF REGULATORY UTILITY COMMISSIONERS

For Immediate Release: **February 15, 2010**

Contact: Rob Thormeyer, 202-898-9382, rthormeyer@naruc.org

NARUC Gas Committee Receives Final Report Detailing

Oil/Gas Moratoria Impacts

WASHINGTON—A public/private Study Group today reported to the nation's State utility regulators in Washington that its research has identified trillions of dollars of impacts resulting from updated domestic oil and gas resource projections and decisions to maintain moratoria restrictions against development of America's oil and gas resources.

The presentation marks the end of an almost two-year study effort administered by the National Association of Regulatory Utility Commissioners and a broad group of public and private interests to determine the financial and environmental impacts of maintaining the moratoria restrictions.

"The previous Administration and Congress removed oil and gas moratoria on public lands over one year ago," Study Group Chairman O'Neal Hamilton said, "but required actions to access the energy resources thought to exist there have not been taken." Hamilton is past Chairman of the South Carolina Public Service Commission and Chairman of the National Association of Regulatory Utility Commissioners' NARUC Committee on Gas, which initiated the study in 2007. "Whether additional Federal lands should be leased for energy development—and under what conditions leasing should occur—is a matter for national energy policy decision makers," Hamilton said. "Our research allows policy makers to know the extent of the resource base and the effects that maintaining the restrictions would have on the country. Our public interest work is dedicated to giving decision makers information upon which they can rely in developing America's national energy policy."

In July 2007, NARUC's Gas Committee and Board of Directors adopted a resolution authorizing the Association's participation in a public-private partnership to study the impact of existing moratoria against energy exploration and production on federal lands, both on-shore and within the 200 mile limit around the Country's coasts.

The "Moratoria Study" used a Federal government modeling program relied upon by Congress and the Administration for analyzing the energy outlook under existing laws and projecting the impacts of new energy policy proposals. The "NARUC-National Energy Modeling System (NARUC-NEMS)" version of the model was employed by the Study Group's contractor, Science Applications International Corporation (SAIC).

Hamilton said SAIC's report provides a comprehensive review and update of domestic oil and natural gas resources, and using the new resource estimates, projects the relative social, economic and environmental effects on the nation of maintaining various moratoria and restrictions on domestic oil and gas exploration and production through 2030.

In updating the resource base, the report estimated that:

- The domestic natural gas resource base on federal lands should be increased over previous estimates by 132 Tcf onshore and by 154 Tcf offshore—a total estimated increase in domestic natural gas from 1748 Tcf to 2034 Tcf. To give that number perspective, in 2009 the United States consumed 22.8 Tcf and, of that amount, imported 2.7 Tcf.

- The domestic crude oil resource base is estimated to increase, offshore, by 37 Bbo (excluding parts of Alaska); onshore, the crude oil resource base is estimated to increase by 6 Bbo for the Arctic National Wildlife Refuge (ANWR), with no estimated increase in the Lower-48 resource base—a total estimated increase in domestic oil from 186 Bbo to 229 Bbo. In 2009 the United States consumed 5.2 Bbo oil, produced 1.9 Bpo at home and imported 3.3 Bbo.

The report projects effects on the nation of maintaining moratoria under two scenarios: 1) determining the social, economic and environmental effects of maintaining moratoria and the updated oil and natural gas resource base estimates, and 2) determining those effects without the updated resource base.

SAIC's NEMS-NARUC model results determined that maintaining traditional energy exploration and production moratoria on Federal lands would result in an alternative domestic energy future that, " . . . increases the cost and restricts the availability of domestic oil products and natural gas . . ." in all economic sectors and regions of the country. According to the study, under the "Combined Comparative Case", which combines the estimated increase in the oil and gas resource base with maintaining moratoria from 2009-2030, model projections show that:

- Cumulative domestic oil and natural gas production decreases by 15% and 9%, respectively.
- Average natural gas price increases by 17% and average electricity prices increase by 5%.
- Cumulative national real disposable income decreases by $1.163 Trillion ($4,500 per capita).
- GDP decreases cumulatively by $2.36 Trillion ($1.16 Trillion NPV), an average annual decrease of 0.52%.
- Cumulative oil imports from OPEC countries increase by 4.1 Billion barrels.
- Cumulative national payments to OPEC countries increase by $607 Billion ($295 Billion NPV).
- If resources within the moratoria areas are not developed, there would be no new environmental effects within the U.S. jurisdiction attributable to development of those resources. However, as a non-modeled item, the study observes that there could be

environmental effects on domestic and international waters as a result of increased tanker transport of oil and gas imports and unknown environmental effects in countries from which the U.S. would import the resources.

An Executive Summary and full report are on the NARUC Website at: http://www.naruc.org.

*   *   *

NARUC is a non-profit organization founded in 1889 whose members include the governmental agencies that are engaged in the regulation of utilities and carriers in the fifty States, the District of Columbia, Puerto Rico and the Virgin Islands. NARUC' member agencies regulate telecommunications, energy, and water utilities. NARU represents the interests of State public utility commissions before the three branches of the Federal government.

# # # #

Edwards Brothers, Inc.
Thorofare, NJ USA
January 27, 2012